T0139314

PLANT ECOPHYSIOLOGY

Plant Ecophysiology

Jean-Claude Leclerc

Professor of Plant Biology and Physiology
Jean Monnet University
Saint-Étienne
France

Science Publishers, Inc.

Enfield (NH)   Plymouth, UK

© 2003, Copyright reserved

SCIENCE PUBLISHERS, INC.
Post Office Box 699
Enfield, New Hampshire 03748
United States of America

Internet site: *http://www.scipub.net*

sales@scipub.net (marketing department)
editor@scipub.net (editorial department)
info@scipub.net for all other enquiries)

ISBN 1-57808-247-1

Published by arrangement with l'Université de Saint-Étienne, Saint-Étienne.

Ourage publié avec le concours du Ministère français chargé de la culture-Centre national du livre.
(This work has been published with the help of the French Ministere de la Culture- Centre national du livre.)

Library of Congress Cataloging-in-Publication Data

Leclerc, Jean Claude,
    Plant ecophysiology / Jean-Claude Leclerc.
       p. cm
    Includes bibliographical references (p.).
    ISBN 1-57808-247-1
    1. Plant ecophysiology. 1. Title.

QK905 L43 2002
581.7--dc 21

2002036501

Translation of: *Écophysiologie végétale*. French edition: © l'Université de Saint-Étienne, Saint-Étienne, 1999. ISBN 2-86272-163-8

Published by Science Publishers, Inc., Enfield, NH, USA
Printed in India

2-T5-1.5 PM

Contents

Table of Figures (Provisional)

Chapter 3: Hydromineral Equilibrium

Chapter 4: Ecophysiology of Development

Chapter 5: Plants in Stressful Conditions

Table of Tables (Provisional)

Chapter 3: Hydromineral Equilibrium

Chapter 4: Ecophysiology of Development

Chapter 5: Plants under Stress Conditions

Introduction

Ecophysiology is the science of interactions between organisms and their environments. It describes the responses of organisms under prevailing conditions and provides a causal analysis of physiological mechanisms and the corresponding ecological factors, at each level of organization.

Ecophysiology can also broadly be defined as the study of causal relationships (or mechanisms) between organisms and their environment, but is it that simple?

The physiological aspects are examined primarily with respect to their adaptive significance. For this, the ecophysiological approach must take into account the polymorphism of individual responses, which are largely responsible for adaptive capacity in any population. Fundamental ecophysiology includes the study of mechanisms underlying adaptive strategies.

Ecophysiology pertains not only to organisms in their natural environment, but also those in conditions manipulated or even created by human intervention.

What is essential is a detailed knowledge of the environment through observation and experimentation as well as the taking into account of periodic variations, and any factor acting in association with the physiological rhythm. The research applies not only to overall responses to variations in a complex environment, in natural conditions, but also to isolated effects of each of the factors involved and their interactions; for study of those aspects, laboratory experimentation is on the whole more appropriate. Thus, there must be a systematic effort to coordinate and relate laboratory experiments and field observations.

Ecophysiology provides principles that explain ecological data and contribute to the identification of mechanisms of dysfunction in ecosystems (Block and Vannier, 1994). At its centre is the concept of adaptation to the environment at different levels of the organism as well as of the medium in which it lives. It is useful to divide those levels into sub-organism (internal) and supra-organism (external) levels. In view of the specialization of researchers at these different levels, there seems to be little and superficial interaction between different research groups. There is a wide range of reviews and there are published books on ecophysiology, but there is no review specifically devoted to this science, except perhaps *Functional Ecology* and the *Bulletin de la Société d'Ecophysiologie*. Perhaps one of the weaknesses of earlier research has been its inability to properly link the results on individuals and species to higher and lower levels.

The growing specialization of research has constrained it within the traditional frontiers of animal and plant biology. Is it possible, in light of the available publications, that researchers of one discipline have often been largely unaware of what has been done in the other?

One question remains: Is ecophysiology simply the study of adaptation to the environment? Terrestrial environments are increasingly rapidly altered by human intervention. New factors are thus intervening, and others are being modified. In consequence, biocoenoses are being deregulated and biotopes are being profoundly changed. New pressures are acting on animal and plant evolution. The ecological approach to these modifications is essentially descriptive; only ecophysiology can provide a satisfyingly explanatory basis.

Ecophysiology can provide the bases for implementation of controlled modifications and also define optimal or at least useful characters for the creation of artificial environments (e.g., extraterrestrial environments) where interactions between living things must occur over long periods.

According to Vannier (1983), ecophysiology occupies a frontier between physiology and evolution. He gives the following definition: Ecophysiology is the science involving the interaction between organisms and their environments. It comprises the descriptive study of the responses of organisms (isolated or in groups) to prevailing conditions and the causal analysis of corresponding physiological mechanisms that are ecologically dependent, at every level of organization.

It is a good definition overall, but what about the evolutionary aspects?

Sibly and Calow (1986) state that strategies of resource allocation are essential to physiology and examine the theory of evolution of life, natural history, and the ecology of behaviour as basic mechanisms of ecophysiology.

According to Bennet (1987), ecophysiology aims to elucidate how animals are "destined" with reference to the natural environment and the history of evolution, historic evolution. But can one explain the evolution of complex groups of organisms simply by ecophysiology? Calow (1987) felt that the function was a matter of processes rather than rightfully a property.

The demonstration of a physiological adaptation is not an end in itself (Feder, 1987). In an evolutionary perspective, physiological characteristics have properties that are not particularly ordinary because they are relatively plastic and can show simultaneously very specific responses and responses of rapid acclimatization to changes in the environment of the organism. In this context they are vastly different from morphological characteristics, which are much slower to respond.

The distribution and extent of physiological adaptations must be observed with care. For example, it can be seen in the works of Clarke (1991) on polar fishes that temperature compensation is not always perfect and that respiration is a particularly misleading index of temperature compensation because it in fact represents the sum of many phenomena, each having a specific reaction with respect to the temperature. Thus, the use of respiration as an index should not be trusted. Similarly, the use of the superfusion point to test the endurance of invertebrates to cold is open to criticism (Block, 1991).

Some authors feel that ecophysiology could become a "service science" for the dominant fields of biology and be relegated to a study of the functional significance of physiological traits relying on the wider studies of molecular biologists or evolutionists (Prosser, 1986). This could lead to a situation in which our discipline will continue to be criticized for showing that animals live where they can (Feder and Black, 1991), while the question is rather how animals are capable of occupying different niches and habitats in different environments.

This question brings us to ask whether animals show specializations in their functional capacities, which are adaptive, or whether there is rather a poor relevancy between the animal and its environment. There seems to be ultimately a combination of nearly perfect and imperfect adaptations in nature, but we cannot assume that all the physiological parameters that we measure are necessarily adaptive. Another way of progressing from this situation has been suggested by Sibly and Calow (1986): one moves from a comparative approach based on correlations (arguments) towards experimental programming and modelling, emerging with a proof of predictions derived from models. Their arguments demand a fresh examination.

The *a posteriori* approach is to assume that all traits are adaptive in one way or another. Such explanations of ecophysiological results are reinforced if the traits observed change in separate populations of the same species placed in different ecological conditions. They respond to ecological shifts and are not fixed, like most traits, which have taxonomic characters. Such comparative and correlative techniques are fundamental to the *a posteriori* approach but they have two major limitations: First, a particular correlation may well have nothing to do with an adaptive cause. Second, the cause may be adaptive but it is not adaptive to that which has been specified at the beginning. As all scientists know, correlations between variables do not prove a cause-and-effect relationship between them. However, rigorous experimental methods will generate confidence in such correlations, and that applies more specially to an ecophysiological study.

The *a priori* approach, by identification at the outset of possible causal factors and formulation of hypotheses to be tested, is often more difficult.

Sibly and Calow (1986) use models for a wide range of organisms and situations and apply a theory of optimality to the science of physiology to explain the ecological processes.

In the modern world we observe phenomena of disorder or desynchronization of biological rhythms that could be understood by ecophysiology. With plants and animals, ecophysiological studies contribute to the optimization of agricultural techniques for the purpose of improving nutrition in terms of quality and quantity. In these fields of research, close links are evident between fundamental and applied research.

With all these experimental aspects, research in ecophysiology could eventually help resolve problems that concern the human race: conservation of natural flora and fauna, improvement of our food sources, a better quality of life, and perhaps our very survival.

Chapter 1

The Environment of Plants

Two major aspects of the environment must be considered: the non-biological environment, which constitutes the biotope, at various scales, and the biological environment, consisting of other plants and animals, which will yield plant-plant interactions and plant-animal interactions. In this section, the biotope will be discussed, along with the major kinds of animal communities according to their typical interactions with plants.

1. THE BIOTOPES

1.1. The atmosphere

The troposphere, and to a lesser extent the stratosphere, have a direct influence on life. All life occurs in the lower part of the atmosphere, which is called the troposphere and is characterized by a continuous fall in temperature as the altitude increases (often –0.4 to –0.75°C/100 m, sometimes –0.2 to –1°C/100 m). The troposphere (Fig. 1.1) is thickest at the equator (16–18 km) and measures 10 to 12 km in the temperate latitudes. It can be divided into two zones, a turbulent lower layer and upper zones of atmospheric calm.

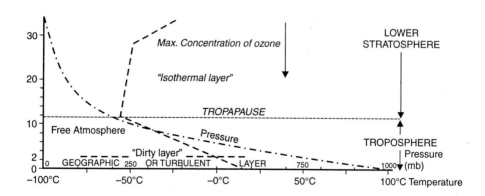

The figure shows a cross-section of the first 30 km of atmosphere

Fig. 1.1. Structure of the atmosphere *(Estienne and Godard, 1979).*

The turbulent layer is around 3 km thick below 45° latitude and encloses most of the water vapour. It is characterized by phenomena of significant atmospheric turbulence due to the plant cover, which causes friction, and to reliefs that cause dynamic ascendances. The interaction with soil and the ocean surface also contributes the creation of large air masses that collide and bring about disturbances of large dimensions. More local interactions bring storms and cyclones. It is in this turbulent layer that most above-ground life exists; vertical movements (rise, subsidence) here are significant, and there are sometimes inversions of temperature (Fig. 1.2) (at a certain height, temperatures increase with altitude) causing zones of very calm air, which could become charged with dust particles and/or mist and clouds, and could constitute "pollution domes" or even, at altitudes of 1 to 3 km, "dirty layers". Thus, the lower layers of the atmosphere are charged with particles that keep out some part of the sun's radiation. These particles ultimately fall to the soil, where they

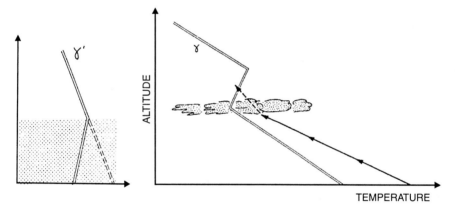

Fig. 1.2. Thermal inversions *(Estienne and Godard, 1979)*.

Table 1.1. Atmospheric deposits in a forest (in mEq/m²/year) (Lindberg et al., 1986)

	SO^2_4	NO_3^-	H^+	NH_4^+	Ca^{2+}	K^+
Precipitation	70 ± 5	20 ± 2	69 ± 5	12 ± 1	12 ± 2	0.9 ± 0.1
Dry deposits						
Small particles						
(< 2 mm)	7 ± 2	0.1 ± 0.02	2.2 ± 0.9	3.6 ± 1.3	1.0 ± 0.2	0.1 ± 0.05
Large particles	19 ± 2	8.3 ± 0.8	0.5 ± 0.2	0.8 ± 0.3	30 ± 3	1.2 ± 0.2
Aerosols	62 ± 7	26 ± 4	85 ± 8	1.3	0	0
Total deposits	160 ± 9	54 ± 4	160 ± 9	18 ± 2	43 ± 4	2.2 ± 0.3

A mixed forest in the eastern United States was studied for 2 years. The number of observations was variable (from 15 for HNO_3 to 730 for SO_2). The overall uncertainty on wet precipitation is around 20%; that of dry deposits rises from 50 to 75%. High precipitation of SO_4^{--} and H^+ is noted because of industrial activities. Also, there are large amounts of SO_2 and HNO_3 deposits in the form of aerosols. The particulate calcium and precipitation of NH_4^+ or NH_3 are not negligible and are useful for the trees.

contribute to the mineral nutrition of plants (Table 1.1). The falling particles may travel long distances (Fig. 1.3) and are effectively captured by forests and mountains.

Most of the water vapour is located in this turbulent layer, as well as a good part of the CO_2, the concentration of which was around 360 ppm in the mid-1990s and is still increasing annually (Fig. 1.4).

Above 3 km there is a "free atmosphere" zone, clearer and more homogeneous, giving place mostly to horizontal movement of air. The temperature drops very regularly, around –0.6°C for every additional 100 m of altitude. Water vapour is rare here, and the atmosphere is dry. The atmosphere is rarefied: 615 mbar at 4000 m as compared with 1015 mbar at 0 m. There is thus little oxygen and little CO_2; animals and plants face a difficult life here and show characteristic adaptations.

At the tropopause (the limit with the stratosphere), the temperature is very low, and this is the limit to which cloud formations can reach. In certain places there are jet streams.

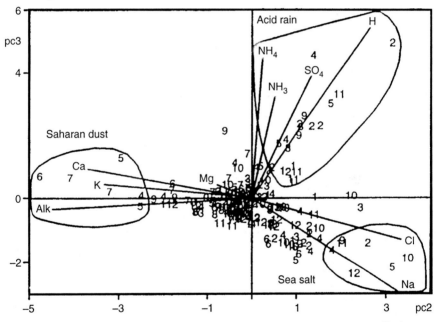

In the forests of Mont Lozere, located at about 1200 to 1400 m, the major contributions are SO_4^{--}, then Ca^{++}, Cl^-, and Na^+. An analysis of principal components was done on the pluvial inputs. The graph shows the second component (20% of the variation) as compared to the third component (17% of the variation). The figures indicate the month in which data were collected. Three types of inputs are clearly seen: of marine origin, mainly in winter; acid rain, mainly in winter; and wind-borne particles from the Sahara, in the spring and summer, bringing useful inputs of calcium and potassium as well as general alkalinity.

Fig. 1.3. Contributions of the wind to a forest in the South Central Massif
(Durand et al., 1992).

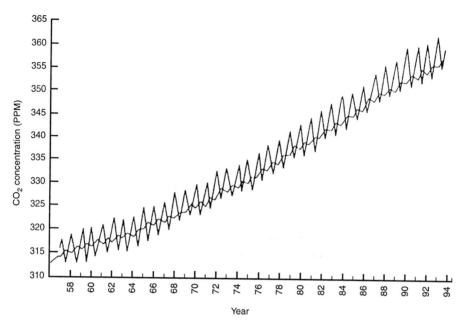

CO_2 was measured in one of the Howaiian islands. As the measure far from any contineut and therefore in full oceanic in, a good mean of CO_2 was obtained. The recorded annual wawes are due to the inbalance of primary product tivity in the Northern and Southern hemispheres.

Fig. 1.4. CO_2 increase in the earth's atmosphere *(Berger, 1994)*

The stratosphere is outside the zone in which life is possible but has an impact on it, particularly by means of photochemical reactions that create or destroy natural or man-made substances by the greenhouse effect, that is by absorbing the remote infrared rays. It also has an impact due to ozone, creating the ozone layer, which absorbs some ultraviolet radiation. The temperature here is very cold, about –50 to –80°C.

1.2. The hydrosphere

There are two parts to the hydrosphere: continental waters and marine waters. The continental waters, running or still, give rise to a wide range of ecosystems, including conditions of low light or even anoxia. For plants, the nutritive conditions and pH can vary greatly. Sometimes there is significant accumulation of salts, particularly NaCl, Na_2SO_4, or Na_2CO_3, which gives rise to physiological adaptations.

Marine waters are much more abundant (97% of the total) and more extensive in area. They are more significant not only as a medium of life and a medium of significant primary productivity, but also in the short term as an essential thermal reserve for the present climate, and in the future as an important CO_2 sink. By their richness in salts, marine waters call for particular adaptations of hydromineral processes in animals and plants.

As in fresh water, the conditions of illumination are highly variable, according to the turbidity of the water due to the charge of mineral particles (a phenomenon less important than in fresh waters) or organic particles, and of plankton.

With some exceptions, depth is overall the most important factor in the marine environment. Plant life is normally possible in the euphotic zone, going some tens of metres deep. There are some plants that in very clear waters can live at a depth of at least 200 m. Marine waters, like continental waters, may have levels of nitrates and/or phosphates that can limit plant productivity. Because of their slightly alkaline pH, close to 8, they are rich in CO_2 (essentially in the form $NaHCO_3$). CO_2 is rarely a limitation, unlike in some fresh waters with clearly acid pH. Oxygen is poorly soluble in fresh water and even less so in saline water, and it can be very limiting in certain saline continental waters, or even in lakes and rivers with high organic pollution.

Marine as well as continental waters have a tendency to be warmer on the surface than at depth, because of the absorption of solar energy. Winds can ensure a certain blending of masses, causing a slight descent of warm waters, but the turnover occurs only down to a certain depth and therefore, especially in summer, thermoclines (abrupt division between warm and cold water, as one goes deeper) may form in lakes and oceans.

Horizontal currents play an important role in the state of the atmospheric climate in some areas, particularly on the eastern and western coasts of the continents, by heating or cooling of the air by contact. Certain cold currents called *upwellings*, rich in nutritive matter, particularly nitrate and phosphate, have an ascending component and bring a great quantity of primary products to the surface.

1.3. The lithosphere

The lithosphere is represented mainly by the emerged part of the continents. The relief and nature of rocks has an impact on the terrestrial environment.

Relief affects life at several levels. The exposure of slopes has an impact on the incidence of solar rays, the steepness of surrounding slopes affects the characteristics of water bodies, and the height of hills determines the altitudes and thus the temperature of a place. Relief determines erosion and the ease with which soils are established. The steeper the slope, the more difficult it is for a soil to be established. Particular dispositions, for example the hollows or coombes of snow determine mesoclimates and microclimates. Relief also can diminish wind action (by sheltering) or increase it (by exposure).

The nature of rocks is manifested mostly in its overall chemical composition, and in the pH it results in. There are different forms of vegetation on acid and basic soils. Particular mineral characters will bring about either a deficiency or excess of a given element. The colour of the rock affects the local climate. Also, its decomposition and fissuration acts on the formation of soils and on the availability of water.

The interaction of the atmosphere and lithosphere results in the creation of a soil at the interface. The general characteristics of the soil—its water content, its mineral and organic richness, its texture, its thickness—depend on the climate as well as on the nature of rocks, and determine the general structure of the biomes. Conversely, the dynamics of an ecosystem also affect the qualities of soil. The biocoenosis also acts, through its management of water, on the climate of a place and also on a larger scale. Finally, the interactions occur at multiple levels.

2. AQUATIC ENVIRONMENTS

2.1. Introduction

The aquatic environment may be permanent or may change daily (zones with rocking marshes, or intertidal zones), or it may be seasonal with more or less regularity (rivulets and marshes that appear only in the rainy season, for example).

The aquatic environment may be defined in relation to the presence of a large body of water, but that would be a narrow definition. In fact, some plants do not get a continuous supply of water. They live on the moisture supplied by mist and fog, as is the case with desert lichens that are drenched at the end of the night.

The environment within a living thing can also be considered an aquatic environment, whether that organism is aquatic or not. This may apply to algae or bacteria that live in symbiosis with aquatic invertebrates (e.g., green or brown algae in a planaria), as well as to ciliate protozoa in the belly of a ruminant.

We must also consider within the context of ecophysiology the very great diversity of aquatic plants, which implies, of course, a great diversity of physiological adaptations. But these adaptations must be placed within the general evolutionary framework; although some groups (many thallophytes) have always been aquatic, others pass from aquatic life to terrestrial life (Raven, 1995, Fig. 1.5) and in certain cases return to aquatic life, which involves not only adaptation to the present environment but also a certain morphological and physiological "memory" of adaptation to an earlier environment.

Aquatic plants are particularly interesting in their photosynthesis, by their enormous diversity in size—from less than one micron to tens of

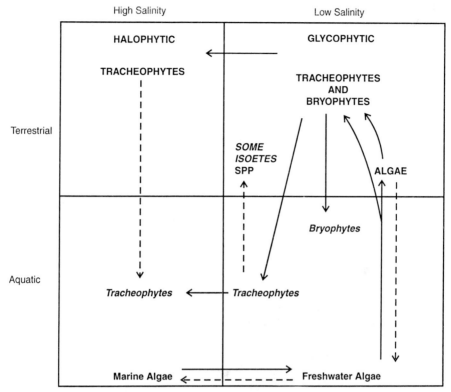

Fig. 1.5. Movement to free air *(Raven, 1995; in Schulze and Caldwell, 1995)*

metres—as well as by the enormous diversity of the light available to them in terms of intensity and spectral quality, and also in terms of variability in time.

2.2. Photosynthesis in the aquatic environment

Photosynthesis in an individual organism depends on the quality and mass of its pigments, which determine the efficiency of energy capture, electron and proton transfer, and finally conversion into organic chemical energy by enzymatic activity.

Aquatic plants often have a particularly complex pigment equipment. There are situations close to those of terrestrial plants, particularly in the Chlorophyceae, and, of course, in the Tracheophytes, which return to aquatic life. Chlorophyll *b*, an accessory pigment in photosynthesis, can capture pale red light and blue light, which extends the possibilities of capturing light by chlorophyll *a* (mostly the medium or dark red or violet); carotenoids are also involved in the capture of energy from violet and blue light. Certain algae with chlorophyll *a* and *b* have carotenoids with particular spectral properties: siphoneins, siphonoxanthins, which extend the capture of photons in the direction of green light.

Many groups have chlorophyll *a* and chlorophyll *c*. Chlorophyll *c* is not very useful for the capture of red light but it is found associated with carotenoids within the chlorophyll-carotenoid-protein complexes, particularly fucoxanthin and peridinine, the last being capable of capturing green light. Certain algae may combine chlorophyll *a*, *b*, and *c*.

A different pigment system exists in the Cyanobacteria, red algae, and Cryptophyceae. It is characterized by the presence of phycobiliproteins, which can be divided into three major groups: the phycoerythrines (PE), which have pics of absorption from 500 to 565 nm, phycocyanines (PC), with pics from 600 to 625 nm, and allophycocyanines (AP) that absorb 650 to 670 nm. All these phycobiliproteins have wide bands of absorption, with a large trail in blue and violet. The energy transfers between them are very good, the bands of fluorescence being wide enough and close enough to bands of absorption of the receptor phycobiliprotein. Finally, the organization of phycobilisomes with a core of allophycocyanines in close contact with the thylakoid membrane allows for an excellent transfer of energy to photochemical centres of photosynthesis. These algae and bacteria thus have the potential to capture all the visible light of the solar spectrum. But we must also take into account that the proteins are extremely costly in terms of nitrogen and energy.

The Cyanobacteria are particularly economical. Many Cyanobacteria live in an aquatic environment in which the quality of light is changing: the light is rather orange when they are close to the surface and rather green at some metres of depth, and also when Chlorophyceae are present at the surface. In these situations, Cyanobacteria show chromatic adaptation (Tandeau de Marsac, 1979). If there is mostly green light, they can make a great deal of PE and only a little PC (for energy transfer); if, on the other hand, there is a great deal of yellow or orange light and little green light (if Chlorophyceae are on the surface), the Cyanobacteria produce a great deal of PC but suppress the synthesis of PE (Tandeau and Marsac, 1979), which would be useless in the circumstances.

The extent of the power of absorption at many wavelengths of visible light is particularly useless for small organisms, which contain little pigment. For a diameter of 0.5 μm, the pigments contained could be distributed over the equivalent of just two to four single-molecule layers. A single-molecule layer of chlorophyll has, at 435 nm, only 0.13 of optical density or 23% of absorption. It may also be noted in this connection that several years ago, in a tropical oceanic environment at a depth of several tens of metres, very small prokaryotes were discovered with a photosystem II that possess divinylic analogues of chlorophyll *a*, *b*, and even *c* (Platt et al., 1983).

All these microorganisms have a significant absorption at all the wavelengths but overall, in individual and not population terms, allow nearly half the light to pass through. Per unit of volume, it is *a priori* useful to increase the concentration of pigments in the cell, and thus the quantity of thylakoids, but, even with an enormous level of thylakoids, for example 40% of dry matter, a radius of 4 μm is required to capture 90% of the light at 435 nm, which is too much for a prokaryote.

If an aquatic organism is large enough, as in the case of several green or brown algae, the spectrum of absorption will have "holes", where pigments absorb little or no light; but the cells being large, where the pigments do absorb, they can capture almost all the light, and if almost 50% of the spectrum is covered with close to 100% absorption, there also the individual will allow almost half the light to pass.

Many vascular aquatic plants have a pigment structure limited to chlorophyll a and b and some carotenoids, but these pigments can be highly concentrated, which by definition allows the plants, by the lateral parts far from the absorption bands, at least partly to fill in the "holes" in the overall absorption spectra. But, in this case, there is the problem of the ratio of cost of investment to the organic matter or energy benefit. The investment may appear very heavy. The same problem is found with sciaphilic plants of forests, especially in tropical environments.

A particular trait of photosynthetic plankton is that it is mobile. Many species are flagellate and can move by turning around an axis of rotation that is also the axis of movement. Suppose the cell is large enough and moves perpendicularly to the direction of the light. Any part of the photosynthetic apparatus will then be illuminated by intermittent light and for a fraction of a second will be illuminated and then again in shadow. During the period of shadow, dark reactions may occur, as reported in the first experiments of Emerson. Thus, the cellular movement may allow optimal use of light energy, and avoid the reception of too much light energy in a single place. This brings us to the phenomenon of photoinhibition.

Photoinhibition occurs when the energy captured by the photosystem II centres is not converted into emission and flowing of "energetic" electrons. The energy of excited chlorophyll cannot flow by normal means and is used to form chlorophyll in the triplet state. This highly reactive chlorophyll, on contact with molecular oxygen, forms singular oxygen $^1O_2$, which is extremely reactive and oxidizes membrane lipids, as well as proteins of photosystem II such as D1.

This results in an inactivation of the electron transport chain, which could, if recurrent, bring about photo-oxidation of the chlorophyll itself. How the ravages of photo-oxidation can be avoided is a question that has been studied mostly with respect to terrestrial plants in a state of stress. For aquatic plants, as for all plants, it is primarily a problem of investment cost versus benefit (Raven, 1995). The most often made investment is the synthesis of carotenoids having more than 11 double liaisons. These carotenoids, which could be different from those of the "xanthophyll cycle" of higher plants, have the property of having a level of excitation that allows them to compete effectively with oxygen for the capture of energy of the triplet chlorophyll, the level of energy being lower in passing from carotenoids to excited carotenoids than in passing from $^3O_2$ to $^1O_2$. The excited carotenoids are deactivated by emission of heat.

A portion of the carotenoids of algae thus has a double function: either to capture energy to achieve photosynthesis (like an antenna) or to avoid photo-oxidation (like a lightning conductor). The benefit is that the organism avoids having to repair the damages of photo-oxidation, particularly by resynthesizing the protein D1.

2.2.1. Form of CO_2 absorbed by aquatic plants

For most aquatic plants, CO_2 is fixed by rubisco. The affinity of rubisco of aquatic plants is often less than that of higher terrestrial plants. Also, it must be kept in mind that in an aquatic environment the diffusion of dissolved CO_2 is around 10,000 times less than that of CO_2 in the atmosphere. Aquatic plants have thus evolved systems that allow them to either concentrate CO_2 in the cells or use biocarbonate (Raven, 1995). In fresh water the pH is often acidic or close to neutral, and there is little bicarbonate but often a great amount of dissolved CO_2, which either comes from sedimentary fermentation or is brought by percolation through soils rich in CO_2. In running water, the limit layer is low and aquatic plants acquire CO_2 without any particular mechanism. The utilization of CO_2 allows them to limit losses of CO_2 from these waters to the atmosphere. Helophytes of marshy zones (*Typha*, for example) rooted in mud rich in CO_2 can circulate the CO_2 by means of aerenchyma and large lacunae in the parenchyma, up to the leaves. Many algae have an active system of CO_2 transport. In fresh or marine water, some species of large marine algae, crassulascean aquatic plants, fix CO_2 by phosphoenol-pyruvate carboxylase, or PEP-carboxykinase, with either metabolic separation in space (C4 metabolism) or separation in time (CAM metabolism). If we extend the observation generally to the marine environment, rich in bicarbonate because of its pH, we see many cases of transformation of HCO_3^- into CO_2, either extracellular or intracellular, by the activity of carbonic anhydrase.

If there is intracellular carbonic anhydrase there is often active transport of bicarbonate. Finally, CO_2 is not likely to be a limiting factor in an aquatic environment; much more often, the limiting factor would be light or nitrogen or phosphorus.

2.2.2. How do plants survive a periodic drying out?

Algae that live in a tidal zone have to solve this problem. In fresh water there may be seasonal drying out and certain mosses and algae such as Rhodophyceae Batrachosperma have to survive such periods. Generally, these species are poikilohydric. For some species a periodic drying out is necessary: this is the case with the Pheophyceae *Pelvetia canaliculata*, and probably also with marine lichens such as *Lichina pygmea*; as for terrestrial lichens, CO_2 fixation is much better after a slight desiccation that removes the liquid film at the surface of the thallus, which screens out CO_2 (Lange, 1989). Some freshwater algae have a diploid perennial stage that resists emersion; the haploid stage (gametophyte) lives only a few months and does not resist it (Raven, 1992). The most general case is that of the large Pheophyceae, which, depending on the level, remain dried out some hours at each low tide. *Fucus spiralis*, which lives in an intertidal environment, realizes nearly a quarter of its photosynthetic production during emersion, which is not negligible. If the alga lives in very turbid waters, the photosynthesis in emersion is still more useful because light is not a limitation if the emersion occurs during the day. These large algae, which absorb useful mineral salts only in water and by the surface of their thallus, must in case of emersion make available some mineral reserves to ensure the physiology of photosynthesis. It can be shown that the absorption of mineral salts is greatly stimulated during the return to immersion.

These large algae must also resist drying out and the excess salinity that may follow at the surface. B. Kloareg and his team studied this question: The large Pheophyceae and Rhodophyceae have particularly thick cell walls around the periphery of the thallus. These walls are rich in alginates, carraghenanes, etc., which have a high gelifying property and are also polyelectrolytes. The walls can thus simultaneously be a hydric buffer and also retain Na^+ ions, an excess of which can also cause damage. It also seems that certain species could capture atmospheric water vapour, according to analysis of isotopic discrimination. Here is a set of interesting questions that demand much more research and clarification.

In conclusion, it can be said that aquatic plants have overall effectively resolved the problems of capturing light energy and capturing CO_2, phytoplankton and the large algae having presented a wide variety of solutions to the first and the aquatic cormophytes having shown diverse as well as effective solutions to the second.

In the context of the use of aquatic plants to limit the increase in terrestrial CO_2, it is necessary to look into the problems of mineral nutrition.

3. SOILS

A soil can be defined as a multilayer granular composite that develops from a mother rock. It is constituted of mineral matter that comes from the decomposition of the mother rock and organic matter from the decomposition of dead tissues and organisms. The two processes of decomposition are in continuous interaction and reach an equilibrium that results in a soil of a particular thickness and overall composition. The resulting soil is influenced by the rainfall and wind, which introduce a mechanical constraint on the overall equilibrium.

3.1. Physicochemical properties of soils

A soil is determined by the nature of the mother rock, the climate, and the kind of vegetation and fauna that are established on it.

The mother rock is subjected to weathering (breaking down) by the climate and living things and the result is debris specific to the original rock, in chemical nature and structure. A primary group of rocks, eruptive rock, is formed from the cooling of very hot matter and leads systematically to the formation of plutonic rock if the cooling is slow and thus forms visible crystals, giving it a granular structure. Examples are granites and diorites. These rocks are most often acidic, as with granites, or base, as with gabbros. They are rich in Al and Si, and their decomposition provides at least a granular fraction. The extrusive (volcanic) rocks are formed by rapid cooling and are thus made up either of fine crystals or amorphous material: e.g., basalts, phonolites, tufas. They may be rich in Ca, and base as with basalts or sometimes acidic as with phonolites. A second group is represented by metamorphic rocks that result from subsidence of parts of the earth's crust due to tectonic movements, which subject them to high pressures and temperatures. Examples of such rocks are marbles, schists, and quartzites. The metamorphic rocks can also come from the encounter of very hot plutonic rocks and schists forming gneiss, mica-schists, and so on. Many of these rocks are very easily broken up. A third category consists of sedimentary rocks that result from the compaction of deposits, from rivers, lakes, or sea water, of biological origin (extracellular structures of carbonates or silica) or resulting from decomposition or erosion of rocks. These result in chalks, clays, sandstone, and conglomerates.

The decomposition, alteration, fragmentation, and breaking up of these rocks under the impact of water, ice, and living things leads to the formation of hard debris that is coarse or fine and colloids such as those of aluminium polyhydroxides. For example, granitic arenite or clays of decalcification may form. Many of these crystals, amorphous granules, and colloids carry negative charges on their surfaces and can therefore retain metallic ions. They thus hold a small or large mass of free water through hydrogen links or by capillary action. All this constitutes the mineral part of soil; it is not in itself a soil but only a rock resulting from the decomposition of other rocks. A soil must also have an organic component.

The organic part is the humus, which comprises substances with a long or short lifetime depending on their resistance to agents of decomposition. Lignins and cutins are molecules directly resulting from living matter. Other substances are products of chemical transformation of relatively resistant parts of matter that was once living, such as the lignins already mentioned. The products of transformation may be water-soluble fulvic acids that have a rather short lifespan (about a year), colloidal humic acids that last longer, and humins that are highly polymerized and last about a century. There are two major types of humus: moder and mull. Moder is relatively acidic. Both kinds are rich in C (45–58%) as well as in O (42–46%) and H (6–8%). They differ in their levels of nitrogen, which range from 0.5 to about 4%, with consequences for quality and stability. Almost all humus comes from the slow or quick decomposition of litter (essentially leaf debris). The annual relative rate of decomposition (in %) can be calculated as k = L/L + T, L being the quantity decomposed each year and T being the total quantity of litter still present.

3.1.1. Litter

In a humid tropical forest, the litter is thin, even though a large number of dead leaves fall on it each year. Since the temperature and the humidity are high, decomposition is very quick. k is about 95%. In a mountain beech grove in western Europe, or even in a conifer forest in western Siberia (taiga), the litter is very thick, because even though the humidity is high the temperature limits the rate of decomposition, and more acicular leaves, as well as the beech leaves, to some extent, are not easily attacked by bacteria, fungi, and microfauna. Thus, there is a fairly low relative rate of annual decomposition, about 10 to 30%.

The result of the decomposition of litter is pedogenesis, that is, an incorporation of organic matter into the mineral fraction of the soil. The cellulolytic bacteria (which hydrolyse cellulose) and nitrifying bacteria (which oxidize the ammoniac equivalent resulting from amino acids into nitrate) trigger the process of decomposition. Depending on the type of vegetation and the climate, there may ultimately be three types of organic formation. Mor humus has a C/N ratio of >30 and is often produced in a coniferous forest, on crystalline rock, and in a cold and humid climate. It develops slowly. Mull humus has a C/N ratio between 10 and 15. It appears in a deciduous forest in a temperate climate, and it rapidly evolves towards a stable humus. If there is much rain, in a cold climate it may form peat, which is more acidic and poorer in nitrogen than mor humus. The formation of peat requires anoxia and low water circulation.

3.1.2. Soil evolution

Soil is not a static medium. It evolves not only over time but also over space. The mineral and organic elements migrate downward and upward, often undergoing chemical modifications. In a hot and dry climate, or a climate that has a long dry season (Mediterranean, tropical), the elements migrate upwards, drawn by evaporating water. If the climate is cold and rainy, the elements migrate downwards by leaching, as for example with the formation of podsols. Humic acids are drawn downwards, accompanied by reaction with clayey particles and the removal of their iron, then they accumulate at depths by releasing iron, which migrates a little further down.

3.1.3. Zonation of soil

As seen with the formation of podzol, a soil ultimately reaches an equilibrium resulting mainly from the climate and the vegetation, and at that point it develops a certain number of zones. In this context, soils are distinguished as zonal, azonal, and intrazonal.

Among the zonal soils there are: permafrosts (soils frozen at depth) or mountain rankers without downward migration; podzols with mor humus and high leaching, thus considerable downward migration; brown soils, often on chalk and in a moderately humid temperate climate, with mull humus and moderate leaching, forming an important B horizon of accumulation, rich in iron; chernozem, with very thick humus (sometimes more than a metre) and no leaching, the rainfall being limited; red Mediterranean soils rich in Fe_2O_3 coming from clays of decalcification; and ferralitic soils, tropical, with very high leaching and also with ascending movement when this soil is bare, which leads to lateritization.

Azonal soils are not yet in equilibrium (for example if they are very young) or never will be because of active erosion. They are often very thin, as in the case of thin rendzina in some karstic soils. The soils of moraines have a great deal to evolve. Cultivated soils are also to be mentioned here.

Intrazonal soils are those that do not have normal zones because of particular local factors, very often excess of fresh or saline water. This results in soils with high saline content (solonetz), or gley soils or gleyified soils in which anoxia will send iron ions into the ferrous state, bringing about the presence of more or less green clays.

3.1.4. Soil texture

The texture of soil depends on its granulometry. The structure depends on the texture and behaviour of certain particles, the hygrometry, pH, and ionic equilibrium of the soil.

Granulometry is the distribution of soil particles into size classes from stones and rocks of greater than 2 cm diameter to clay particles of less than 2 μ diameter. Granulometry is fundamental to the water retention capacity of soils, oxygenation, and propagation of heat and cold (thermal conductivity), which is important for water supply and root physiology. The structure depends on the granulometry and behaviour of at least some of the particles. Fine colloidal particles may flocculate and form aggregates that cement the larger elements, which will lead to the formation of lacunae for the circulation of air and water, and also give some stability to the soil in terms of overall physical qualities. If, on the contrary, the colloids remain dispersed, the soil particles remain independent. Then there is no defined lacunar system and we have a more or less unstable particulate soil.

3.2. Hygrometry of soils

The water retention capacity of soil depends heavily on its porosity and the available surface area of grains, which increases as the grain size becomes smaller. The soil porosity may be defined as the lacunary volume/total volume.

After a heavy rain, the water fills all the lacunary volume; the soil reaches its maximum retention capacity A; in the hours that follow, the water in the largest cavities and between the largest grains will percolate downwards and represents a very free water with water potential (ψH_2O) close to zero. This is the gravitational water B. What remains in the soil is water fixed essentially by capillarity, the ψH_2O of which is around 0.1 atm. This water represents the field capacity, C = A – B.

An important fraction of C represents the available water (from ψH_2O = 0.1 to ψH_2O = 8 to 15 atm), i.e., the water that can be used by plants. A little capillary water, between the finest particles, cannot be extracted by plants and can be lost by air drying. Finally, there is water of hygroscopicity.

Hygrometry and granulometry give rise to particular physiological and morphological adaptations of root systems (Fig. 1.6).

3.3. pH and ions

Soil pH is greatly influenced by the mother rock. The composition of leaf litter also plays a special role through its C/N ratio. Often, abundance of Ca^{++} is correlated with a neutral or slightly base pH. The Mg^{++} also increases the pH.

The micelles of clayey particles and humic compounds retain a large amount of cations on negative surface charges. This results in a cation exchange capacity (CEC) establishing a buffering power. The CEC is saturated if there are 100% of metallic cations fixed on the particles. It is zero if there are 100% of protons, i.e., complete desaturation.

Much K^{++} is fixed on the micelles; it comes mostly from feldspaths and mica (thus from crystalline rocks). The Mg^{++} comes from deeper rocks: olivine, amphiboles, pyroxene; serpentine, which is particularly rich in it, leads to the presence of special plant ecotypes (edaphic ecotypes) on the soils that develop from it.

4. REACTIONS OF PLANTS TO BIOTOPES

4.1. General characteristics vis-à-vis various parameters

The various characters of a biotope act directly on the possibility that a particular plant will live in a given place. Soil humidity and pH have a

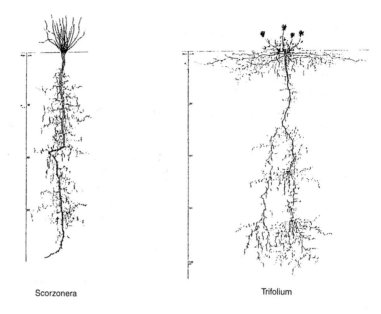

Scorzonera   Trifolium

Allorhizic roots

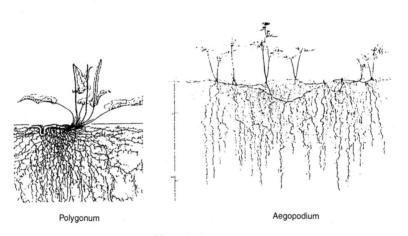

Polygonum   Aegopodium

Homorhizic roots

Some plants have a main root or taproot, for example *Scorzonera villosa* (a species of oyster-plant) or *Trifolium trichocephalum*, which also has a network of lateral roots at the surface. In more evenly distributed root systems, many roots of equal size radiate from the base of an above-ground rosette or rhizome (*Polygonum bistorta*, a knotgrass found in fairly moist mountainous prairies with leaves that can be used in soup) or emerge from underground stolons as in *Aegopodium podagraria* (ground elder), which often grows in the undergrowth of riverside woods.

Fig. 1.6. Major types of root systems (*Kutscher and Lichtenegger, 1992 in Larcher, 1995*).

decisive influence. Other important characters are illumination, temperature with respect to the regional average, altitude (which is linked to the temperature), and the presence of nitrates, calcium, and other minerals. Some of these characters, such as illumination, are themselves influenced by the plants, particularly the dominant plants. Relatively similar species from a taxonomic point of view often have similar biotopes but this is not always the case. Finally, some species can live in many environments, and others have strict needs with respect to the value of certain parameters, such as pH.

All this can be illustrated with some examples.

- **Humidity**

Ribes rubrum (garden red currant) is hygrocline, i.e., it has a preference (cline) for rather humid environments; *Ribes nigrum* (black currant) is mesohygrophilic, i.e., it requires a highly and constantly humid environment, with frequent flooding. In France, the other species of this genus have a better range of tolerance of dryness and humidity. *Ribes petraeum* (red currant) can be found in an intermediate situation between dry and humid (mesophilic) but mainly in a humid situation (hygrocline). *Ribes alpinum* (alpine currant) is found in slightly humid to slightly dry environments, and *R. uva-crispa* (gooseberry) is found in intermediate environments (mesophilic) sometimes very windy and exposed, or in slightly humid or humid environments. This last species has a much larger scope of adaptation. Sometimes, even when there is plenty of water, it is not easily available, generally because it is charged with salts, and we then see the presence of more or less strict halophytes (*Salicornia* or *Cochlearia*).

- **pH**

Ribes is never found, for example, in a highly acidic environment. It often prefers environments that are neutral or slightly acidic. Among the heathers, *Erica cinerea* is xeroacidiphilic, i.e., it has a fairly strict requirement of acidity and dryness, but *E. tetralix* is hygrophilic and acidiphilic, preferring humid to very wet terrains that are always acid. *Sorbus aucuparia* (rowan) accepts highly acidic soils but also chalky soils. *Lactuca perennis* (perennial lettuce) prefers dry and chalky soils; it is a calcicolous xerophile. The fern *Athyrium distentifolium* (alpine lady-fern) is neutrocline and hygrocline.

- **Light**

The light that reaches the soil (Fig. 1.7) comprises a small percentage of ultraviolet light (a proportion that increases with altitude) and around 50% of visible light. The rest is infrared. Independently of the nebulosity

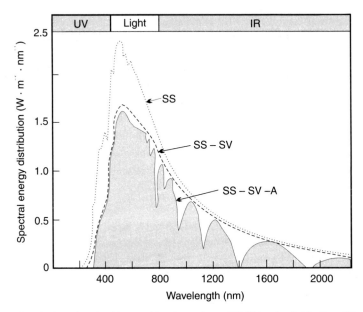

SS, light before being absorbed by earth's atmosphere. SS-SV, solar light after diffusion and absorption by the upper atmospheric layers. SS-SV-A, light that reaches the soil surface. The SS-SV-A curve presents a maximum at 470 nm, if there are no clouds and if radiation is vertical. As the sun sets, the spectrum tends towards the red: at 10° elevation of the sun (around half an hour before setting), λmax = 650 nm. The ozone layer eliminates short-wavelength UV rays (λ < 300 nm). The hollows of infrared emission around 900, 1100, 1400, and 1900 nm correspond to bands of absorption by water vapour up to 1900 nm and beyond that there is also absorption by CO_2.

Fig. 1.7. Spectrum of sunlight *(Schulze and Caldwell, 1995)*

and the nycthemeral period, seasonal variations are increasingly important towards the high latitudes and must be taken into account in the extratropical zones. Part of this light, nearly all the infrared and a part of the visible spectrum, is reflected by the soil and by the vegetation itself (Fig. 1.8).

When the light is low, thus limiting for photosynthesis, morphological adaptations occur in plants of the undergrowth, and these are particularly developed in tropical environments (Fig. 1.9). *Paradisia liliastrum* (St. Bruno's lily) is heliophilic, while *Circea alpina* (enchanter's nightshade) and *Oxalis acetosella* (wood-sorrel) are strictly shade plants. Certain species have a wide range of light requirement, such as *Teucrium scorodonia* (wood sage), which is found in sunny places and in shaded places in a fairly dry forest.

● **Temperature**

There are many specifically alpine plants adapted to very cold winters and cold summer nights, for example, *Androsace alpina* (alpine rock-

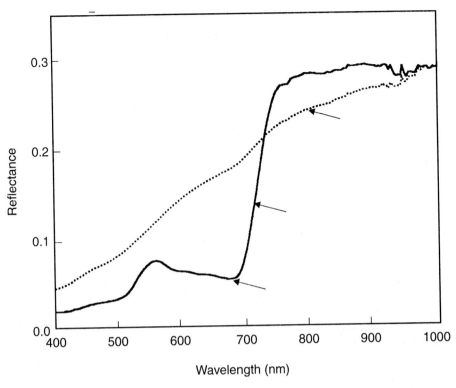

Reflectance is a means of analysing the distribution and nature of the vegetation on a large scale by remote detection. The figure shows the reflectance of a sunflower field; the considerable increase between 700 and 750 nm is typical of plants and allows their general identification.

Fig. 1.8. Reflectance of vegetation and bare soil
(Schulze and Caldwell, 1995).

jasmine) or *Cerastium cerastoides* (starwort mouse-ear). Some trees of the subalpine zone, such as *Pinus cembra* (Swiss stone pine), are particularly resistant to winter cold.

- **Particular mineral characters**

Plants that are nitrophilic have extended their areas considerably due to human activities. *Urtica dioica* (perennial stinging nettle) is a hygrocline nitrophile, *Rumex obtusifolius* (broad-leaved dock) is nitratophile, and *Viola calaminaria* is specific to soils rich in zinc and lead. Sometimes plants living in the same place have different mineral contents (Fig. 1.10).

Beyond a common general physiology, many plants have special physiological mechanisms or certain common traits are particularly well developed, which allows them to live in characteristic, sometimes extreme, environments.

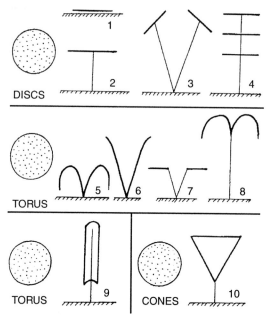

Plants of the tropical forest undergrowth develop leaf and foliage forms designed to avoid suffocation by forest litter, and above all to best exploit the light coming from above as well as the diffused or reflected light they receive laterally when they are near a clearing or the forest edge. The overall forms with axial symmetry are represented here: acauline disc (1), elevated disc (2), disc cupola (3), superimposed discs (4), torus in acauline spray (5), torus in a funnel (6), torus in wreath (7) torus in descending spray (8), cylinders (9), and cones (10).

Fig. 1.9. Overall forms of plants of the tropical forest undergrowth
(Blanc, 1992).

4.2. Plants in normally extreme situations or exceptionally abnormal situations

An extreme situation can be said to correspond to life in a biotope with characteristics perceptibly different from those in which life is considered easy or at least easier. Plants living in extreme biotopes have gradually evolved the necessary genetic and physiological characters, which may in some cases date back to epochs in which the first life forms appeared.

Plants of biotopes considered favourable may be faced more or less suddenly with a situation that could be considered extreme, either for them or in general. In a situation of stress, the plant, usually a higher plant, will react more or less felicitously, depending on the intensity of the stress and the adaptive capacities of its genome.

It can be said that life in an extreme environment corresponds to a permanent stress, but what does that mean? Situations of stress are highly variable, and the reactions of individuals are also highly variable. Stresses may be more or less habitual or they may be exceptional; an exceptional

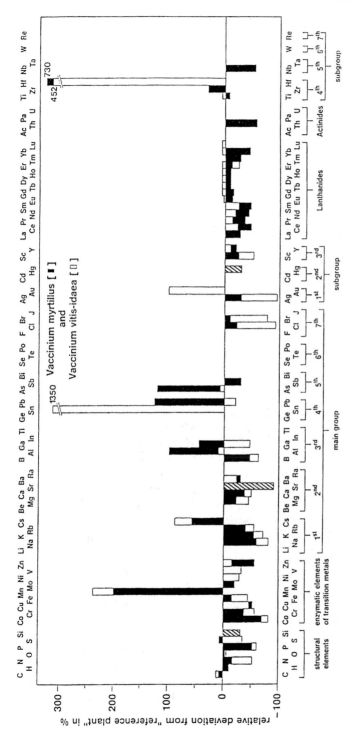

Fig. 1.10. Differences in mineral content of sympatric and taxonomically related species *(Markert, 1994).*

First of all, it is observed by comparison with a reference plant representing the average value of a large number of higher plants that certain elements are particularly well assimilated by two species of *Vaccinium*: for example, manganese, caesium, and hafnium. On the other hand, most of the alkalines (except Cs) and alkaline composts are low, which has to do with the acidophilic character of these plants. There are some marked differences between the two species. *Vaccinium vitis idea* (mountain cranberry) accumulates only tin, and *V. myrtillus* (bilberry) accumulates only lead. Gallium is also treated differently, higher than the average in *V. myrtillus* and lower in *V. vitis idea.*

situation may be not transitory, that is, followed by a return to normal, but irreversible, or at least sufficiently long-lasting to correspond to a large number of generations, which brings us to life in habitually extreme conditions.

Those conditions must be considered at the scale of the individual or part of an individual as well as on the scale of populations, ecosystems, and even biomes.

To come to specifics, we briefly discuss below some extreme conditions as well as the most frequent stress situations.

4.2.1. *Extreme conditions*

Some extreme conditions—drought, cold, heat—may be considered general, i.e., affecting all life regimes. Extreme conditions could involve not just a single factor, but an overall extreme situation: e.g., life at high altitude, life in very polluted environments, or life in a weightless atmosphere. Some conditions are extreme only for some groups of organisms, depending on their trophic modes.

Except for parasites, plants must be able to photosynthesize in their habitual situation. Thus, we can consider life in a cavern (but not far from the entrance) or at a certain depth in the ocean or a lake (but with a minimum of light) to be a life in extreme conditions. Animals and bacteria that live at very great depths and thus under great pressure are also in extreme situations.

How do we define an extreme environment? First, there are two closely linked aspects in such an environment: the number of ecological niches is small and the number of species present is also reduced, i.e., few genomic solutions have been found. Also, an extreme environment usually cannot be used, or can only with great difficulty be used, by a foreign species introduced by humans. In defining extreme environments, we must also ask, extreme for whom? Certain types of living things, the Archebacteria and lichens, for example, can live in extreme environments. Such an environment is not extreme for them but is so for higher plants. Only a few rare species of higher plants can live in it, or sometimes none.

We must also keep in mind that some species have wide ecological amplitudes and their extremes could be considered extremes in themselves; the human species is the most obvious example. The Eskimos, who were well described by Malaurie, and the now extinct Alakaluf Indians who lived in conditions appalling to most Europeans, in the islands close to the Tierra del Fuego, all lived and thrived in environments where the marine life was extremely rich. In the plant kingdom, the snow buttercup is practically the only higher plant species found above 4000 m. In western Europe, this *Ranunculus* is found (although not exclusively) in the northernmost territories of Europe and Siberia; it is also found along with a certain number of lichen species.

An extreme environment can thus be described as one in which few species are found, or only very few representatives of the most highly evolved biological groups, such as higher plants.

4.2.2. Some environments extreme for plants

Some environments are extreme in terms of cold, heat (Tables 1.2, 1.3), pH, wind, and the rarity or inaccessibility of water and light. Also, several factors must be considered jointly, for example light and cold, or even heat and acidity. These problems are discussed in detail in Chapter 5 of this work.

4.2.3. Very hot environments

In a volcanic environment, hot springs and geysers are inhabited even though their surface temperatures reach around 95°C. Also, there are populations of cyanobacteria and sometimes mosses at temperatures from 50 to 70°C, in vents emitting steam.

In a natural situation, or in artificial reliefs and structures, high temperatures can be observed even in a temperate climate, on soils that are very dark and/or soils on a steep slope exposed to the south or southwest. For example, on a mining tip close to Saint-Etienne, a beautiful birch (*Betula pendula*) lives in a southern exposure, on a soil that reaches surface temperatures of over 55°C in summer. In a hot and dry climate, especially in desert or sub-desert subtropical zones, air temperatures of up to 58°C can be observed.

At great depths, near hot springs in rifts in the ocean bed, life can be observed in temperatures of up to 110°C. Life in these extremely hot environments demands a particular type of metabolism.

The archebacteria present a group of extreme thermophiles, and some are also adapted to high acidity or salinity (Pelmont, 1993): *Pyrodictium occultum* shows optimal growth at 105°C.

"Mean" temperatures of optimal growth can be observed with *Sulfolobus solfataricus* (87°C) or *Acidianus infernus*. These germs develop in anaerobic conditions and often use sulphur as a source of energy. They are insensitive to antibiotics that inhibit the protein synthesis of other bacteria. They also have numerous peculiarities in their transfer RNA. In assembly or construction of ribosomes, the details of which are very well known in *Escherichia coli*, we observe with respect to the latter particular hydrophobic interactions that allow it to create an overall stability at high temperature.

In most groups of living things, there seem to be, in case of thermal shocks, a certain number of special proteins that either protect them from denaturation or (in the case of chaperonins) repair the other proteins. Chaperonins to some extent guide and favour restoration, which is spontaneous in many proteins. This allows us to find the exact conforma-

Table 1.2. Temperature limits of microorganisms and poikilohydric plants (Larcher, 1973; Brock, 1978; Fujikawa and Miura, 1986; Stetter et al., 1990, in Larcher, 1995)

Group	Cold (after 2 h application on hydrated organisms), °C	Heat (after 30 min. application), °C	
		Hydrated	Dry
Bacteria			
Archebacteria		100 to 110	
Cyanobacteria and others			
Photosynthetic bacteria		55 to 75	
Saprophytic bacteria		60 to 70	
Thermophilic eubacteria		up to 95	
Bacterial spores		80 to 120	up to 160
Fungi			
Phytopathogens		45 to 70	
Saprophytes	0 to –10	40 to 60 (80)	75 to 100
Fructifications	–5 to –10, sometimes -30		
Spores		50 to 60, sometimes > 100	
Algae		100	
Marine algae			
Tropical zone	5 to 14, sometimes –2	32 to 35, sometimes 40	
Temperate zone			
Infralittoral	–2 to –8	25 to 30	
Littoral	–8 to –40	30 to 35	
Polar zone	–10 to –60	20 to 28, sometimes 15	
Freshwater algae	–5 to –20, sometimes 30	35 to 45	
Subaerial algae	–10 to –30	40 to 50	
Thermal springs	15 to 20	45 to 50	
Lichens			
Polar regions	–80		
High mountains	–80		
Deserts	–80		
Temperate regions	–50	33 to 46	70 to 100
Mosses			
Tropical zones	–1 to –7		
Temperate zones			
Humid habitats	–5 to –15	40 to 45	
Undergrowth	–15 to –25	40 to 50	80 to 95
Epiphytes and epiliths	–15 to –35		
Polar regions	-50 to -80		
Reviviscent ferns	–20	47 to 50	60 to 100
Reviviscent phanerogams			
Ramonda myconi	–9	48	56
Myrothamnus flabellifolia			80

Poikilohydric organisms in the dry state are particularly resistant to cold, often to temperatures of –196°C.

Table 1.3. Resistance limits of some organisms at extremely cold or hot temperatures
(in Larcher, 1995)

Taxonomic group	Resistant to cold (°C)	Resistant to heat (°C)
Bacteria		
Archebacteria (oceanic hot springs)		100 to 110
Thermophilic cyanobacteria and certain		
photosynthetic bacteria		55 to 75
Some thermophilic ex-bacteria		90 to 95
Fungi		
Saprophytes		40 to 80
Phytopathogens		45 to 70
Lichens		
Polar regions	−80	
Deserts and high mountains	−80	
Temperate zones	−50	
Algae		
Subaerial algae	−10 to −30	40 to 50
Ocean and polar glacier algae	−10 to −60	
Algae of thermal springs	15 to 20	45 to 50
Mosses		
Forest soil surfaces (temperate zone)	−15 to −25	40 to 50
Epiphytes and epiliths (temperate zone)	−15 to −35	
Polar regions	−50 to −80	
Polyhydric ferns	−20	47 to 50
phanerogams (50% leaf damage after		
2 h cold or 30 min. heat)		
CAM succulents	−5 to −15	58 to 67
Desert winter annuals	−6 to −10	50 to 55
Cold regions		
Conifers	−40 to −90	44 to 50
Shrubs of arctic and/or alpine zones		−30 to −70

Resistance is indicated for organisms in the hydrated state; it is greater in the dry state for
a whole organism (e.g., lichens) and for the spore state or other resistance form (e.g., gram
positive bacteria, fungi).

tion of origin. What is that conformation in the case of thermophilic
archebacteria?

When a thermal shock is administered to *Pyrodictium* by leaving it at
108°C, a type of proteic particle, already present at 90–100°C, appears in
large quantity (73% of hyaloplasmic soluble proteins, at 108°C). This
particle is an ATPase and it is a chaperonin. Its extraordinary abundance
suggests that it may serve in the same way (and possibly at the cost of an
expense of ATP) to protect the remaining cytoplasmic proteins.

These thermobacteria all use sulphur as a respiratory acceptor (in the
presence of H_2) or as a substrate to be oxidized (in the presence of O_2).

The *Sulfolobus* that live at pH as low as 1 and at an optimum pH of
2.5 require hydrogen and use sulphur as a respiratory acceptor to be able

to assimilate CO_2. The same is true for *Pyrodictum brockii*. *Acidianus infernus*, which grows at 65 to 95°C and at a pH of 1 to 5, lives in aerobic conditions and oxidizes sulphur (substrate) with the help of oxygen (respiratory acceptor) to form sulphuric acid. These two processes allow thermophiles to release the energy necessary to the synthesis of organic matter. Hydrogenase (an enzyme that uses H_2 and dissociates H^+ and electrons from it) of *Pyrodictium brockii* is in itself an essential enzyme and is of course highly thermostable.

4.2.4. Very cold environments

Very cold environments are found particularly in the high mountains, and also at high latitudes on steep slopes where snow does not remain on the

Table 1.4. Adaptation of lichens to cold and heat (Larcher, 1995)

Species and geographic origin	Cold temp. at which assimilation begins	Zone of good photosynthetic activity		Hot temp. at which assimilation ends	Reference
		C50	H50		
Picytonema glabratum (= *Cora pavonia*), Venezuelan Andes, 1770 to 3600 m	0	10	35	40	Larcher and Vaneschi, 1988
Ramalina maciformis, Negev rocks	−6	5	32	37	Lange, 1969
Cladonia alcicornis, sandy terraces of the Rhine in the Alps	−5	−3	18	26	Lange, 1965
Stereaucaulon alpinum, moraines of Central Alps, 2400 m	−10	−5	18	22	Lange, 1965
Usnea aurantiaco, oceanic tundra of					Kappen and Redon, 1987
Antarctic					
erect form	−6	0	25	29	
prostrate	−4	0	18	23	
Usnea sphacelata,	−10	0	22	28	Kappen, 1989
continental Antarctic					
Buellia sp., endolith of rocks in dry valleys of Antarctic	−6	−1	12	16	Kappen and Friedmann, 1983

Lichens were studied in the hydrated state. C50: cold temperature at which half of the maximal rate of photosynthesis was reached. H50: hot temperature at which half of the maximal rate of photosynthesis was reached.

ground. In these areas, lichens develop, as well as some higher plants (Table 1.4). The adaptations may be morphological (plants in cushions, very thick bark) or even metabolic (presence of special proteins, poly-osides). In frost, structures develop so as to limit the formation of ice crystals. This effect of intense cold is accompanied by dryness: one of the components of the resistance of plant cells is the manufacture of small molecules that greatly reduce the water potential and thus effectively reduce its freezing point. Certain hormones such as abscisic acid have an important role in adaptation to cold. Since cold represents one of the most typical and thoroughly studied stresses, it is further discussed in Chapter 5 of this work.

5. THE BIOTIC ENVIRONMENT

5.1. Introduction

Plants, or even a single plant, must live in interaction with all the other living things constituting the biocoenosis. The biocoenosis exerts interactions at different levels. First among these are interactions between major

Table 1.5. Interaction between two populations (Dreux, 1986)

Type of interaction of populations A and B	Value of coefficients p (pop. A) p' (pop. B)		Result of interaction
Neutralism	0	0	No reciprocal action
Competition	< 0	< 0	One population tends to eliminate the other
Mutualism, symbiosis, or cooperation	> 0	> 0	Reciprocal help
Commensalism (A commensal of host B)	> 0	0	Obligatory for A
Amensalism (A amensal, B inhibitory)	< 0	0	A is inhibited, B is not affected
Parasitism (A parasite of host B) or predation (A predator of prey B)	> 0	< 0	Obligatory for A, B is inhibited

y = size of a population A. r = constant characterizing the rate of reproduction.
$dy/dt = ry$ = rate of population growth, from which $dy/y = rdt$ and $1/y \times dy/dt = r$.
In a population that is not (or is no longer) in exponential growth, a factor ky and $1/y \times dy/dt = r - ky$ must be introduced, where $k = r/K$, K being the maximum size attained by the population. If $K = k$, of course $1/y \times dy/dt = 0$, the population is no longer growing and is in a stationary state.
If two species are living together, each influences the population size of the other.
We go from $1/y \times dy/dt = r - ky$ to $1/y \times dy/dt = r - ky + pz$, z being the size of population B of the other species and p having a positive value if the second species has a beneficial effect, and a negative value in the contrary case. Similarly, a coefficient p' is defined for the action of the first species on the second.
Thus, we have: $1/y \times dy/dt = r - ky + pz$ for A and $1/y \times dy/dt = r' - k'z + p'y$ for B.
p and p' are positive, null, or negative values in six main types of interactions.

groups at the trophic level, such as relations between plants and herbivores and relations involving carnivores. Other interactions are more or less specific, for example, parasitism or symbiosis. Relations between two species (Table 1.5), of plants for example, can be modelled. There are also relations between two or more individuals of the same species. Many kinds of competition occur, more or less complex, and end in an equilibrium that makes it possible to produce enormous biomass in the terrestrial (Table 1.6) as well as the aquatic environment (Table 1.7).

Biomass values follow from productivity (Tables 1.8, 1.9), which is expressed per day or per year.

5.2. The environment of animals

Animals prove to be indifferent or active in their effect on plants. Interactions may be positive or negative for the plant. The plant-herbivore relation is generally considered negative for the plant, but it may also have positive aspects. In a prairie, browsing favours the growth of grass. The consumption of fruits may result in spreading of seeds which will favour the dissemination of the plant that has been consumed.

Table 1.6. Some biomass values of terrestrial environments
(from several authors, according to Dreux, 1986)

Ecosystem	Biomass (kg/ha)
Vegetation	
Prairie on laterite in Guinea	200 to 500
Savannah in Senegal	590
Marsh in Michigan	4650
North American prairie	6550
Savannah in Guinea	5000 to 1000
Mangrove in Puerto Rico	63,000
Oak forest in Great Britain	128,000
Oak and beech forest in France	275,000
Primeval forest in Congo	1,000,000
Primeval forest in Amazon	1,700,000
Soil microflora (bacteria, algae, fungi) in	
a cultivated field in Switzerland (0 to 15 m depth)	20,000
Invertebrates (litter and herbaceous stratum) in a:	
Prairie on laterite in Guinea	7 to 25
Sansouire in Camargue	55
Savannah in Guinea	250
Birds	
Heathland in Germany	0.005
Pond edges in Germany	1.3
Mammals	
Sahara	< 0.1
Canadian tundra	8
Forest of Congo near Lake Kivu	235

Table 1.7. Some biomass values in aquatic environments (Dreux, 1986; Ramade, 1990)

Ecosystem	Biomass (kg/ha)
Benthic invertebrates	
Mediterranean Sea	100
Baltic Sea	200
English Channel	400
Bering Sea	1650
North Sea	3460
Antarctic	13,500
Fishes	
Trout lake, USA	60
Eniwetok atoll	446
Dead arms of rivers, USA	625
Plant population	
Oceans (in general)	10 to 50
Continental plateaux	10 to 400
Coral reefs and algal growth	400 to 40,000
Lakes and rivers	up to 1000

Table 1.8. Productivity of aquatic ecosystems (Dreux, 1986; Ramade, 1990)

Ecosystem	Gross (g/m^2/day)	Net (g/m^2/day)
Open sea, Pacific	0.2	
Open sea, Sargasso Sea		
in summer	0.5	0.26
over a year	0.55	0.26
Lake Erie		
in winter	1.0	0.26
in summer	9.1	
Danish coast		
in December		0.01
in August		0.7
Lakes and rivers (general avg)		0.8
Okhostk Sea in May		2.0
Coastal waters, Long Island Sound	3.2	
Continental plateau (general avg)		1.0
Hot springs, Silver Spring, Florida	17.5	7.4
Pacific coral reefs	18.2	around 12

Other interactions are immediately positive for the plant, particularly in the case of pollinating animals, but we must not forget that, to attract them, the plant must produce substances that are a drain on its energy budget, to the point where the dynamics of the individual suffer. Certain animals improve the soil (Fig. 1.11). There are also more complex situations involving plants, herbivores, and carnivores, the plant favouring the life of the carnivore in order to diminish herbivore pressure.

The energy cost for the plant is significant in the framework of its reproductive effort, particularly for the formation of seeds, most of which

Table 1.9. Productivity of terrestrial ecosystems (Dreux, 1986; Ramade, 1990)

Ecosystems	Productivity (net, $g/m^2/year$)
Extreme deserts	3
Nevada desert	40
Low prairies of Wyoming	69
Tundra (avg)	140
Prairies in Nebraska and Oklahoma	446
Deciduous forest in England	1560
Temperate coniferous forest (avg)	1300
Boreal forest or taiga (avg)	800
Tropical savannah (avg)	900
Equatorial forest in Java	5400 to 6900
Marshlands (general world avg)	2000
Population of tropical papyrus	7200
Gallery forest in Thailand	9100
Sugarcane cultivation (world avg)	1725
Agro-ecosystems (general world avg)	650
Sugarcane cultivation (maximum in Hawaii)	6700

Table 1.10. Reproductive effort in undergrowth and seed size (Bierzychudek, 1982)

Species	Avg dry weight (mg)
Allium ursinum	5.4
A. victorialis	6.0
Anemone virginiana	0.7
Arisaema triphyllum	34.4
Cimicifuga racemosa	2.9
Claytonia virginica	1.2-1.3
Dentaria laciniata	1.8
Desmodium glutinosum	0.03
Dicentra canadensis	1.0
D. cucullaria	1.2-2.0
Erythronium albidum	6-9.7
Geranium maculatum	5.6
Geum canadense	0.8
Hieracium venosum	0.3
Hydrophyllum appendiculatum	21.0
Isopyrum biternatum	2.2
Lysimachia quadrifolia	0.8
Mitchella repens	2.3
Sanguinaria canadensis	7.1
Thalictrum clavatum	0.7
T. dioicum	1.7
T. polygamum	1.6

The reproductive effort of herbaceous plants and chamaephytes of the undergrowth is relatively low, often between 2 and 10%. They have, as shown in the table, a significant tendency to make large seeds, by which a young plant can establish itself effectively in a difficult situation, for example, by rapidly making large leaves from reserves. The number of seeds, however, may be small, for example around 5 for *Arisaema triphyllum* (an aroid plant) or 0.5 to 5.8 for *Erythronium albidum*. Many species described are American and have European vicariants.

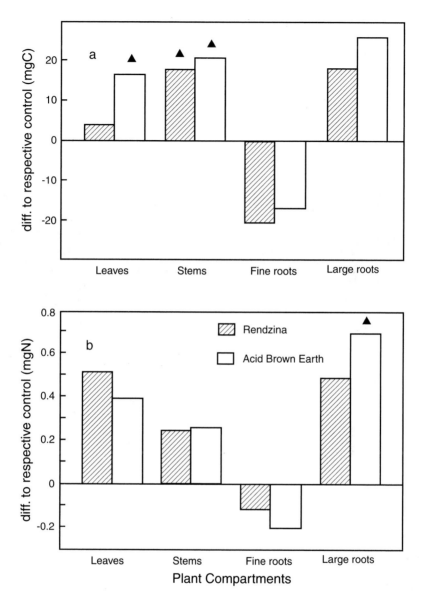

The experiment was conducted in a laboratory with soils (rendzina and acid brown earth) taken from a beech forest; the worm *Octalasion lacteum* Orley and germinated plants aged 4 months were studied over 80 days. Part a of the figure shows the difference of biomass in relation to the respective controls, on leaves, stems, and large roots. Part b shows the assimilation of nitrogen. It is seen that the beneficial effect of worms is more frequent on acid brown earth than on rendzina. Particularly, there is a significant effect (triangles) on biomass of leaves and branches.

Fig. 1.11. Improvement of plant productivity by earthworms
(Wolters and Stickan, 1991).

are sometimes conserved by herbivores. When the plant's energy resources are limited, it must limit its seed production, but it must then protect those seeds (Table 1.10).

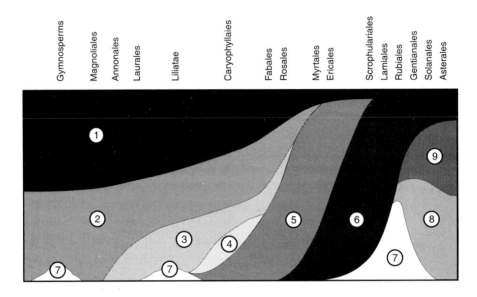

Over the course of evolution, compounds (1 to 5) derived from shikimate first appear, then the derivatives of the acetate and mevalonate route, which partly replace the former. 1. lignin. 2, condensed tannins. 3, isoquinolic alkaloids. 4, betaines. 5, gallic tannins. 6. indolic alkaloids and iridoids. 7. steroids. 8. sesquiterpenes. 9. polyacetylene.

Fig. 1.12. Secondary metabolism in the macroevolution of cormophytes
(Frohne and Jensen, 1992, and Harborne, 1988, in Larcher, 1995).

Table 1.11. Energy cost of some substances (Paine, 1971; Lieth, 1975 in Larcher, 1995)

Substance	Energy content (KJ/g)
Oxalic acid	2.9
Glycocole	8.7
Malic acid	10.0
Pyruvic acid	13.2
Glucose	15.5
Polysaccharides (glucanes)	17.6
Proteins	23.0
Lignin	26.4
Triglycerides	38.9
Terpenes	46.9

Note the energy cost of terpenes, which are used as such or after modification as a signal or defence chemical.

A plant defends itself against herbivores by evasive measures, mechanical means, and chemical means. Chemical defence is the result of a

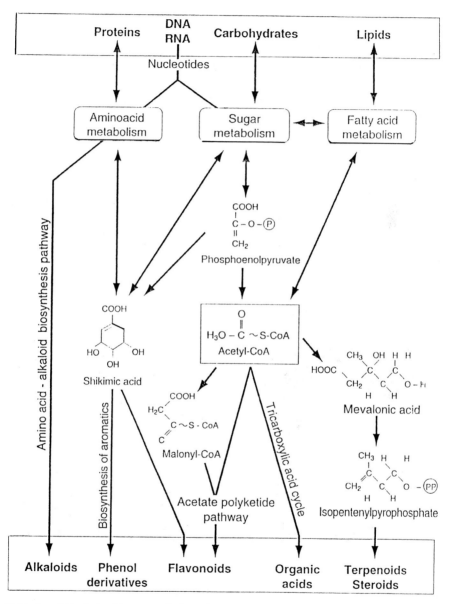

Shikimic acid is derived from metabolism of sugars and amino acids. Essentially from sugars and fatty acids, acetyl-CoA leads to the formation of malonyl-CoA and mevalonic acid.

Fig. 1.13. Major routes of secondary metabolism
(Schlee, 1992 in Larcher, 1995).

long evolution that has made use of secondary metabolites (Fig. 1.12) that are increasingly adapted but also involve high energy cost (Table 1.11 and Fig. 1.13). The plant thus produces toxic, repellent, and non-appetizing substances.

5.3. The role of microorganisms

Most types of microorganisms interact with plants. Soil microorganisms in particular play a much more important role than was earlier believed. Diazotrophic symbiosis, established between *Rhizobium* and the members of the family Fabaceae, has long been known, as well as symbiosis between Actinomycetes and certain wood species. Mycorrhizal symbiosis has also long been part of scientific landscaping and forestry studies. In some cases mycorrhizae have been threatened by pollution (Fig. 1.14).

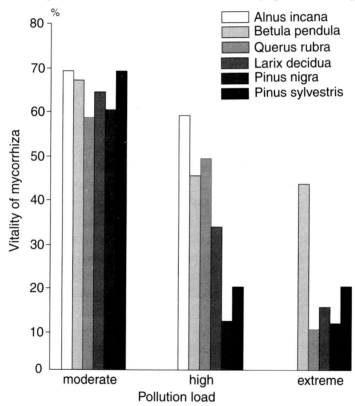

Note the resistance of birch mycorrhizae to extreme pollution, whereas those of alder are eliminated. Oak mycorrhizae show fairly good resistance to serious pollution. *Alnus incana*: white alder. *Betula pendula*: weeping birch. *Quercus rubra*: red oak. *Larix decidua*: European larch. *Pinus nigra*: Austrian pine. *Pinus sylvestris*: Scotch pine.

Fig. 1.14. Vitality of mycorrhizae in industrially polluted zones, mostly acidogenic *(Kowalski, 1987 in Larcher, 1995)*.

Many soil bacteria and some soil fungi are pathogens. Other factors act positively at the level of the **rhizosphere**, particularly among cultivated plants.

Many plant-microorganism interactions use the aerial route, which allows dissemination of pathogenic bacteria and fungi over long distances.

Many of these interactions are also complex, and often animals serve as vectors of pathogenic species. For example, aphids serve as vectors of viral diseases. Sometimes bacteria will fight herbivorous insects, as with *Bacillus thuringiensis*, which is used (or its toxins are used) in modern biotechnology.

5.4. Effects of other plants

Gardeners have long observed that certain trees—e.g., the walnut, several *Prunus*—or certain shrubs such as thyme have an adverse effect on the

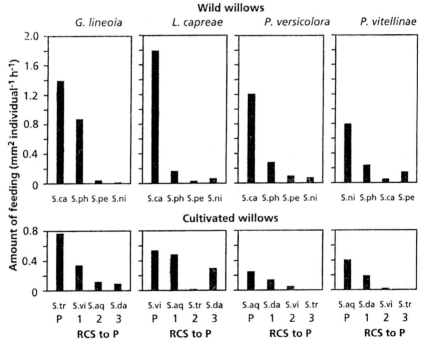

Each species of Coleoptera is habitually associated with a species of willow. The Coleoptera are *Galerucella lineola*, *Lochmaea capreae*, *Plogiodera versicolora*, and *Pratora vitellinae*. Three of them feed mostly on *Salix caprea* (S. ca) and one on *Salix nigricans*. The non-indigenous willow species were not preferred.

Fig. 1.15. Choice of nutrition for four species of Coleoptera among
several species of willow
(Tahvanainen et al., 1985 in Lambers et al., 1998).

Table 1.12. Comparison of some fields of Ericaceae: dynamics and plant-animal interactions (Mallik, 1995)

Character	*Calluna* field	*Kalmia* field	*Gaultheria* field
Origin of field	Fire, forest cutting	Fire, forest cutting	Fire, forest cutting
Climate	Oceanic, humid, temperate	Oceanic, humid, temperate	Oceanic, humid, temperate
Geographic distribution	Western Europe	Eastern Canada and USA	Western Canada and northwestern Pacific
Landscape	Small shrubs more or less prostrate, incursions in undergrowth	Small shrubs more or less prostrate, also in light undergrowth	Large shrubs and also in undergrowth
Hospitability	Hostile to *Picea sitchensis* (sitka spruce), which ceases to grow	Hostile to *Picea mariana* (black spruce)	Hostile to *Tsuga plicata, Thuja heterophylla, Picea sitchensis, Pseudotsuga menziessi* (Douglas fir)
Phenology	Perennial and evergreen	Perennial and evergreen	Perennial and evergreen
Regeneration	Mostly vegetative, but also a good seed reserve in soil	Mostly vegetative, also a good seed reserve in the soil	Mostly vegetative, also a good seed reserve in the soil
Growth phase	Cyclic in improved and managed lowlands	Continuous, no apparent phases	Continuous, no apparent phases
Palatability (acceptance by cattle)	Good value, palatable to sheep and grouse	Little value for cattle, not palatable to sheep and grouse	Good pasture for wild herbivores

subsequent cultivation of certain plants. Botanists and biogeographers have long sought to explain the disposition of species among themselves in ecosystems as well as the existence of zones in which certain species grow in exclusive populations, eliminating every other species from their neighbourhood. Gradually the notion of allelopathy was conceived, a phenomenon by which an individual or a group eliminates other plants from its surroundings through chemical warfare.

There are also plants that compete for the use of resources such as water, nutritive ions, and light. In many cases, allelopathic and trophic pressures are combined.

The effects of the biocoenosis are only briefly presented here. They are analysed more thoroughly in Chapter 5, which covers the ecophysiology of stress. Such effects can be studied in ecosystems that seem relatively simple (Table 1.12), or in controlled conditions with fewer species (Fig. 1.15).

Chapter 2

Photosynthesis and Plant Production

1. INTRODUCTION

Photosynthesis, a major route for the uptake of energy, is the basis for all life functions. It appeared very early on earth through photosynthetic bacteria, since the atmosphere and water did not provide organic molecules or minerals yielding a large quantity of energy, or in sufficient concentrations, except in particular ecosystems. Photosynthesis may be considered to have developed in two major stages. First there was formation of a photochemical system with bacterial chlorophyll, which, when it absorbs light energy, can achieve a **separation of charges** that allows electron activation from substances (still sufficiently abundant in the terrestrial atmosphere) having Em_7 values that could be considered averages, for example H_2S ($S/HS^- = -0.18$ V), H_2 ($H_2/2H^+ + 2e = -0.43$ V (but in fact the values are closer to zero in the natural environment, because of pH and low concentration), succinic acid (close to -0.1 V), and fumaric acid (close to 0 V).

Electrons activated by light energy increase their potential to about -0.5 to -0.6 V. This allows reduction of several iron-sulphur proteins, more or less linked to the membrane, such as ferredoxin, which can, in turn, reduce NAD or NADP ($Em_7 = -0.32$ V). Such reactions provide all the reductive energy at the cell level. $NADHH^+$ and NAD^+ have a difference of free energy of around 50 kcal/mole.

The activation of electrons also allows, through the crossing of the membrane, the creation of a membrane potential. Finally, the membrane quinones, which are intermediary transporters of the electrons, also transport H^+ from one side of the membrane to the other. Thus, in linkage with CF_0 and CF_1 complexes, a photophosphorylation takes place, i.e., a synthesis of ATP exploiting the proto-motor force created by the light energy. This can also be illustrated in a diagram of the photosynthetic membrane of purple non-sulphur bacteria or green non-sulphur bacteria (Fig. 2.1).

Instead of using organic acids, or even H_2, photosynthetic bacteria on a similar schema (Pelmont, 1994) use H_2S, which was abundant in the

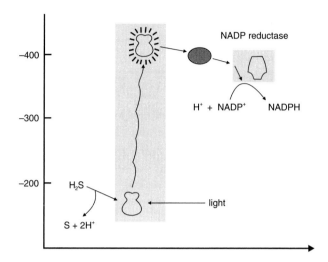

Photosynthesis of a green sulphur bacteria is represented here with, on the X-axis, the sense of movement of electrons, and, on the Y-axis, the oxidation potential (in mv). Fd = ferredoxin.

Fig. 2.1. Photosynthesis of a single photosystem
(*Albers et al., 1983*).

primitive terrestrial atmosphere. This still holds for purple sulphur bacteria (e.g., *Chromatium vinosum*) or green sulphur bacteria (e.g., *Chlorobium limicola*) with more or less complicated modalities.

Thus, what are conventionally called the photosynthetic bacteria probably began to diversify and increase the world of living things from 3500 million to 2500 million years ago.

2. PHOTOSYNTHESIS WITH TWO PHOTOSYSTEMS

2.1. Introduction

Once the easily usable reducing molecules in the oceans were exhausted, living things had to find other means to produce energy. About 2000 million to 2500 million years ago, certain rocks with iron in the ferric state appeared, which indicated the existence of at least a small amount of free oxygen in the air and water. This oxygen came from the oxidation of water, which released electrons in immense quantity but with great difficulty (Em_7 = +0.81 V). Photosynthetic bacteria could not oxidize water. It was the Cyanobacteria that did it by means of a second photochemical reaction (the reaction of photosystem II), which allowed electrons to be transferred to the old photosystem (I).

Cyanobacterial photosynthesis is the same as the photosynthesis found in algae and cormophytes (Fig. 2.2), barring in a few details. It is

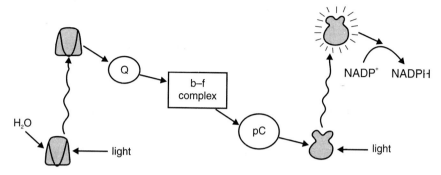

PLANT CHLOROPLASTS AND CYANOBACTERIA

At right, photosystem I. At left, photosystem II. Q, quinones (including plastoquinone). PC, plastocyanin, which transports the electrons between the two photosystems across the lumen.

Fig. 2.2. Photosynthesis with two photosystems
(Albers et al., 1983).

strongly subjected to environmental pressures, particularly light, to the extent that ecophysiology of photosynthesis can be considered a science in itself. During the transfer of energy across the different stages of photosynthesis we will examine the impact of various environmental factors.

2.2. Photochemical reactions

In the beginning, light is absorbed by a chlorophyll molecule, within about a femtosecond. This is not the molecule that triggers the activation of the electron, but one of many others that represent the **chlorophyll antenna** and, with possibly other pigments, form part of the collecting antenna of energy. It is necessary for the plant to capture most of the energy that reaches it. Thus, plants show adaptations at all levels of their structural organization, leaving as few "holes" as possible by which light can escape.

This chapter is limited to the description of those adaptations. Universally, part of the **collecting antenna** is made up of chlorophyll antennae that are fixed on proteic chains; also, there are always two proteic complexes (one for each photosystem) that carry some chlorophyll antennae, and a special "dimer" of chlorophyll that is capable of converting the energy received by the chlorophyll antennae into electron energy due to the emission of an electron: these are the **reaction centres**.

In the case of reaction centres of photosystem II (Fig. 2.3), the activated electron is received by a primary acceptor of the electrons, a pheophytine, which becomes a reduced pheophytine, within 10^{-7} seconds.

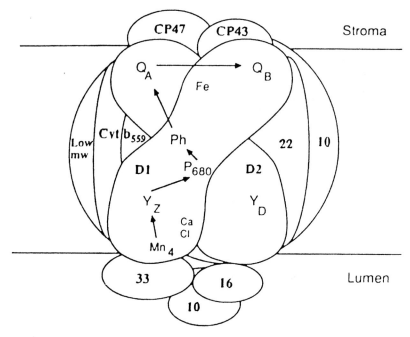

Many transmembrane proteins carry prosthetic groups. Protein D1 shows the reaction centre P_{680}, pheophytine (Ph), iron, and Q_B (plastoquinone coupled with peptide); then there is a group of manganeses, the functioning of which is stabilized by Ca^{++} and Cl^- as well as by several peptides that are in contact with the lumen, which gives an electron at Y_Z that is the final donor of the electron at P_{680}. Y_Z is represented by the amino acid tyrosine 161. Protein D2, which is quite symmetrical to D1, apparently has much fewer functions than D1. Here there is the electron acceptor Q_A (another fixed plastoquinone) and also an electron donor Y_D. On the side of the stroma are observed chlorophyll-protein antenna complexes CP43 and CP47, which are fixed. Other chlorophyll-protein antenna complexes may after phosphorylation move towards intergrain spaces (and the other way around) to balance the inputs of light energy of the two photosystems.

Fig. 2.3. The membrane complex of photosystem II
(Trebst, 1995, in Schulze and Caldwell, 1995).

The chlorophyll antenna effectively absorbs red as well as violet and blue light, but it is ineffective in absorbing the rest of the spectrum. To absorb nearly all light, plants can do two possible things: either accumulate chlorophyll in great quantities (see section 3.1 of this chapter) or use other captor pigments, which will absorb the blue-green, green, yellow, and orange. These pigments are called **accessory pigments**.

The accessory pigments may be mainly true carotenes and carotenoids (mostly in algae) or even phycobiliproteins in red algae and cyanobacteria (see Chapter 1, section 2.2). For the captured energy to be transferred effectively up to the photochemical proteic complex, the pigments must be very close to one another on the proteins or in proximity to them; the chlorophyll proteins also must themselves be close together.

A strict structural organization is thus necessary. If there are problems, some or much of the energy may be lost in the form of fluorescence before it arrives at the photochemical centre.

Accumulation of chlorophyll, accessory pigments, and organization of complexes are closely determined by the light available in the plant's environment.

When the energy finally (in 10^{-9} seconds) arrives at the chlorophyll pair of the photochemical centre called P_{680}, the chlorophyll pair passes into the excited state, then an electron is expelled and positive chlorophyll is formed: chl.chl$^+$. The electron subsequently reduces pheophytine.

The proper functioning of the chlorophyll pair and of its primary acceptor in the plant is itself linked to environmental factors: excess light, drought, salinity, industrial pollutants, and other factors that may disturb their functioning. If the Chl-Chl pair cannot liberate an electron it loses its energy in the form of fluorescence. The measurement of fluorescence (see section 4) is an essential tool in assessing the state and productivity of the entire system.

2.3. Movement of electrons

The electron energy, as in bacteria, crosses the membrane and fixes itself on the quinone proteins (comparable to those of bacteria), then circulates in the pool of plastoquinones. At this level, various herbicides, industrial pollutants, and the temperature will greatly intervene in its functioning (see section 4).

Electrons subsequently pass into the cytochrome f b_6 complex; at this level the plastoquinones becoming oxidized release a large quantity of H^+, which goes into the lumen. They are not the only ones. It must be recalled that when the Chl-Chl pair of photosystem II loses an activated electron, it must return to its fundamental state in capturing an electron coming from elsewhere, i.e., water, which is produced by the intermediary of a cycle of reactions of an enzymatic character allowing the capture of 4 electrons and release of a molecular oxygen from two molecules of water. Five steps follow during the course of this cycle with liberation of $4H^+$ in the lumen of the thylakoid.

The proteins of the **donor system** of the electrons in photosystem II are highly sensitive to the presence of calcium and manganese, manganese being essential by virtue of its levels of oxidation to different electron transfers that occur between water and the Chl.Chl$^+$. The mineral nutrition of the plant plays an important role at this level (see section 3.3).

The protons resulting from decomposition of water and those coming from oxidation of the plastoquinone by the intermediary of a cyt f b_6 complex construct a very steep gradient of pH between the luminal domain and the hyaloplasmic domain of the chloroplast. This pH gradient provides the essence of the **proton-motive force** of the chloroplast which permits photophosphorylation.

As for the electrons, when they leave the cyt f b_6 complex, they are taken in charge by a water-soluble transporter: the plastocyanin, which is a protein containing copper.

Plastocyanin migrates in the lumen, going from membrane zones rich in photosystem II, and situated mostly in the granular structures of the chloroplast, towards intergranular membrane zones, which mostly carry photosystems I. The respective importance of granular and intergranular membrane zones depends greatly on the luminous environment in general as well as the nychthemeral cycle.

There is a considerable analogy of function, and partly of structure and organization, between the complexes of reaction centres of photosystems I and II. In conditions of high acidity of lumen, plastocyanin has a redox potential of around +0.4 V, coming close to a complex reaction centre of photosystem I. It can donate an electron if the Chl-Chl pair is oxidized.

Photosystem I receives energy according to the same processes as photosystem II. It arrives at the core of the reaction system on a dimer of chlorophyll that realizes the separation of charges, the P_{700}, the redox potential of which is +0.45 V. The formation of Chl-Chl* followed by Chl.Chl+ allows the release of an activated electron up to a potential that is about −0.6 to −0.7 V.

This electron crosses part of the membrane in order to fix itself on ISP: intrinsic proteins very rich in iron and sulphur. At this level, the importance of the mineral nutrition will be noted. The electron subsequently passes on a particular extrinsic ISP: ferredoxin ($Em_7 = -0.42$ V). This gives the charge to an extrinsic enzymatic complex, NADP oxido-reductase, permitting the formation of a good reducer of chloroplast: the NADPHH+ ($Em_7 = -0.32$ V).

At this level the chloroplast thus possesses NADPHH+ and ATP, which in the presence of CO_2 will allow the synthesis of all the organic matter that a plant needs. The overall reaction of this non-cyclical transport of electrons involving the two photosystems can be expressed as follows:

$$2H_2O + 2NADP^+ + 2ADP + 2Pi \xrightarrow[P_{680} + P_{700}]{8\ photons} 2NADPHH^+ + 2ATP + O_2$$

The quantitative yield of nb photons/nb O_2 emitted is generally about 0.05 to 0.12. In passing, the importance of mineral phosphate uptake is noted.

2.4. Biochemical reactions

2.4.1. Fundamental reactions

Organic matter, or **photosynthates**, are synthesized by the Calvin cycle, in the stroma of the chloroplast and with the participation of numerous intermediate sugars and a series of enzymes, several of which are **activated by light**. The Calvin cycle "begins" with the fixation of CO_2 on ribulose 1.5 bisphosphate with the help of the enzyme that is most abundant on the earth and represents around half the soluble leaf proteins: ribulose 1.5 bisphosphate carboxylase oxygenase or **Rubisco**. By this means two molecules of phosphoglyceric acid (PGA) are produced, which using $NADPHH^+$ are reduced into phosphoglyceric aldehyde, a sugar in C_3 that is the source of other sugars. Phosphorylations intervene and at each turn of the cycle the equivalent of $1/6$ of a molecule of fructose 1.6 bisphosphate is produced. The metabolism of CO_2 fixation using only this Rubisco is called C3 metabolism. The overall reaction can be expressed as follows, showing the utilization of ATP and $NADPHH^+$:

$$nCO_2 + 2nNADPHH^+ + 2nATP \xrightarrow{\text{enzymes}} (CH_2O)n + 2nNADP^+ + 2nADP + 2nPi + nH_2O$$

The input of CO_2 into the system is, however, essential. Its intracellular concentration regulates the overall rate of formation of photosynthates. The intracellular concentration depends on the external concentration in water, or on partial pressure in the air. CO_2 is also in more or less rapid equilibrium with HCO_3^-. Carbonic anhydrase is an extremely important enzyme of the walls of algae and of plant chloroplasts that greatly enhances the rate of $HCO_3^- \rightleftarrows CO_2 + OH^-$ interconversion and thus the liberation of CO_2 for Rubisco, if the bicarbonate is abundant (neutral or alkaline environment).

Carbonic anhydrase is an enzyme with zinc (here the importance of mineral nutrition is evident), which, with processes of acidification based on the ATPase activity, is at the basis of various mechanisms of CO_2 concentration. This problem is important because the enzyme Rubisco has a relatively low **affinity** to CO_2. In fact, the rate of photosynthesis is directly linked to the internal intercellular concentration of CO_2 in the leaf (Fig. 2.4), which depends on the resistance to diffusion (Fig. 2.5).

Through the ages, CO_2 concentration in the terrestrial atmosphere has decreased and some plants have improved their utilization of CO_2 at low concentration by means of a more refined enzyme that leads to CAM and C4 metabolism.

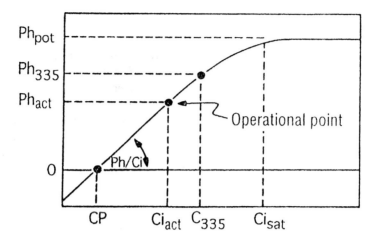

Leaf internal CO_2 concentration

The potential photosynthetic capacity can be reached, Ph_{pot}, which is the maximum possible (at a given temperature) if the CO_2 concentration around the cells reaches Ci_{sat}. When the stomata are fully open, and in normal air, an effective intercellular concentration is reached, Ci_{act}, and an effective photosynthetic activity, Ph_{act}. This is the "operational point". Ci_{act} is inferior (by about 100 vpm) to Ca or concentration of the ambient air (which is actually around 355 vpm) because of the stomatal resistance. The range between Ci_{act} and C_{335} increases (as it does between Ph_{355} and Ph_{act}) if the stomata close. A Ph/Ci slope can be defined that indicates the photosynthetic utilization of CO_2. Ph/Ci divides the line of null net photosynthesis at CP, compensation point of CO_2.

Fig. 2.4. Relationship between photosynthesis and the intercellular CO_2 internal concentration (Ci) in the leaves
(Lange et al., 1987).

2.4.2. CAM and C4 metabolism

2.4.2.1. Crassulacean acid metabolism

Some plants are adapted to a very hot and dry environment. In these conditions, a plant of C3 type loses a great deal of water if it develops through photosynthesis. All plants must conduct gaseous exchanges with the external environment by the intermediary of epidermal structures of the leaves or stems called the stomata.

Stomata can regulate their opening (see Chapter 3) to allow the entry of CO_2 in the plant according to a gradient. The problem is that the opening of stomata also allows water to escape from within the plant. Inside the plant, there is saturation of water vapour, and outside the plant the partial pressure of water vapour is low, and thus there is a steep gradient.

If a C3 plant is in a dry, hot environment, it must often be irrigated to compensate for its high water loss. In a natural, non-irrigated environment, it must close its stomata and have thick epidermal cuticles (see

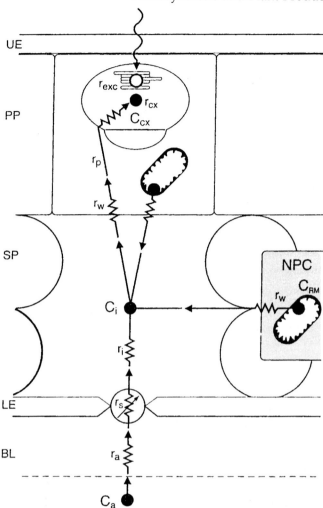

During photosynthesis a CO_2 gradient is established between the free air (outside the outer layer) where the concentration is maximal (C_a) and the chloroplast stroma of the palisade parenchyma, where it is minimal (C_{ox}). The CO_2 also comes from mitochondria, where it is at a concentration C_{RM}, and from photorespiration. A resistance of the outer layer r_a is observed that causes a primary fall in concentration at the entry into stomata that opposes a **variable resistance** r_s to the entry of CO_2. Then there is a resistance r_i due to the passage of CO_2 into the lacunae between the cells, and a resistance r_w due to the passage into the cell walls (saturated with water). Finally, there is cytoplasmic resistance (r_p) and resistance of carboxylation (r_{cx}) at the level of the stroma. In C3 plants, photosynthesis depends ultimately on C_i. C_i may be inferior to C_a to the extent that the air is calm (large boundary layer) and the stomata are closed. UE: upper epidermis. PP: palisade parenchyma. SP, spongy parenchyma. LE, lower epidermis with stomata. BL, boundary layer.

Fig. 2.5. Resistance to movements of CO_2 and resulting concentrations
from the exterior to the interior of the plant
(chiefly Chartier and Bethenod, 1987; Larcher, 1995).

Chapter 3) to avoid losing water, but then it can no longer photosynthesize effectively.

How is this problem solved? Can the plant develop by means of photosynthesis without losing too much water when water is scarce? Yes, by means of crassulacean acid metabolism (CAM).

The plant closes its stomata during the day and opens them only at night. Since the night in a desert environment is quite cold, the succulent plant—examples are members of the families Agavaceae, Cactaceae, or Euphorbiaceae—loses very little water and can fix CO_2. It has been seen that Rubisco has little affinity for CO_2: the internal CO_2 concentration is thus always quite high, the gradient is also fairly weak, and the entry of CO_2 is slow. Moreover, the CO_2 is fixed on a phosphate sugar, with formation of PGA, thus of phosphate molecule. In order to accumulate PGA, it would be necessary to have an enormous quantity of phosphate.

During the night, Rubisco is not used, but another enzyme, phosphoenolpyruvate carboxylase or PEPcase. PEPcase has two advantages. First, it has a great affinity for CO_2. Even when there are only a few ppm of CO_2, the enzyme is saturated, and there is thus a high gradient and rapid fixation. Second, a product of carboxylation is obtained: a dicarboxylic acid called oxaloacetic acid, which can be transformed into malic acid.

Malic acid can be accumulated in large quantity in the vacuoles of cells of a succulent plant. When the sun rises the plant will have some photosynthesis using Rubisco alone (Fig. 2.6) but it must very quickly close its stomata. All the other exchanges must, in this closed situation, occur within the plant: malic acid must go out of the vacuoles, be lysed by malic enzyme, and release CO_2 within the plant, and thus at high concentration.

This CO_2 can be effectively fixed by Rubisco, and the Calvin cycle synthesizes sugars, with the help of an input of ATP resulting from photophosphorylation, and an input of $NADHH^+$ resulting from electron transfers from photosynthesis. The pyruvate produced by malic enzyme will be stored and transformed into phosphoenolpyruvate, which will be used during the following night.

CAM plants thus function by realizing two carboxylation reactions that are distant in time, one during the day and the other during the night.

Another metabolic strategy in hot environments is C4 metabolism.

2.4.2.2. C4 metabolism

For a long time, it has been noticed that certain cultivated tropical plants, particularly members of Poaceae (sugarcane, maize, sorghum), have remarkable growth and productivity of glucides, in a hot environment that is not constantly humid. Examination of the leaves of these plants, as in

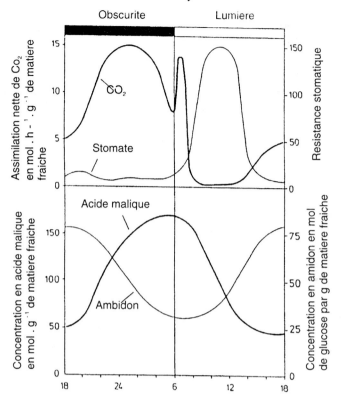

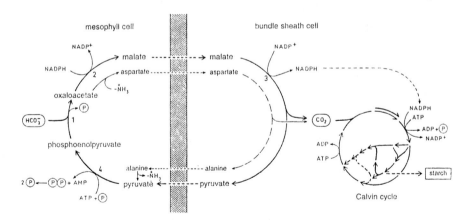

A: stomatal resistance and assimilation of CO_2 (above) and levels of amidon and malic acid (below). B: diagram of C4 metabolism and relations between a mesophyll cell and bundle sheath cell.

Fig. 2.6. Photosynthesis of CAM and C4 plants
(Lüttge et al., 1992).

the leaves of other Poaceae, shows the presence of vascular cords. But these cords are surrounded by a mantle of large cells with relatively clear content, in which there are chloroplasts enclosing relatively little chlorophyll. These particular cells constitute the **perivascular sheath**. The chloroplasts show little granular stacking, and the photosynthetic function must therefore differ from that of other chlorophyll cells of the foliar parenchyma, called cells of the **mesophyll**.

When photosynthesis of these plants with a typical perivascular sheath is studied, it is seen that the curve of photosynthetic activity as a function of CO_2 concentration differs from that of C3 plants. The C4 plants have a very low point of CO_2 compensation, just a few ppm of CO_2.

Analysis of enzymatic content shows presence of enzymes of the Calvin cycle, including Rubisco and also a great quantity of PEPcase. It is also observed that Rubisco activity is present in sheath cells, but PEPcase activity is present only in the mesophyll cells. Unlike with CAM plants, the two enzymatic activities are not separated in time (they both occur during the day), but they are separated in space (Fig. 2.6).

The CO_2 is fixed by subepidermal mesophyll cells. By means of PEPcase and carbonic anhydrase, it forms oxaloacetate, which can be derived from two routes: transamination with formation of aspartate, which is directed towards the mitochondria or, more significantly, towards the chloroplasts. The chloroplasts of the mesophyll reduce oxaloacetate into malate. The malate is directed, by the symplastic route, essentially towards the sheath cells, where it is hydrolysed by malic enzyme into pyruvate and CO_2 with the restitution of $NADPHH^+$.

The CO_2 then easily enters the Calvin cycle, and it ultimately forms saccharose or amidon. The formation of concentrated CO_2 is at the basis of a very high productivity of C4 plants (Table 2.1), and these plants are also observed to have a relatively low transpiration in relation to the productivity. The water use efficiency (WUE) or transpiration quotient can be calculated (Table 2.2). The WUE is high for CAM plants (since they open their stomata only at night) and still higher for C4 plants. The advantage of C4 plants can be explained as follows: If the opening of stomata is identical to that of a C3 plant, the respective water vapour gradients will be roughly the same, and the losses of water vapour are equivalent. But the CO_2 gradients are nearly twice as high in the C4 plants because the partial tensions of CO_2 in the substomatal chambers are very different, and the CO_2 flow, thus initiating the photosynthetic activity, twice as great. It is the low internal pressure of the CO_2 in the cells of C4 plants that explains a WUE that is twice as high as in C3 plants. The rates of photosynthesis of C4 plants may be very high (Table 2.1); C4 plants are also less sensitive to oxygen (Table 2.3) and have very high levels of light saturation (Table 2.4).

Table 2.1. Maximum rates of net photosynthesis under normal conditions of CO_2, temperature, and humidity in plants known for their productivity and in particular groups (data from numerous authors, particularly Bjorkman et al., 1972; Seeley and Kammereck, 1977; Patterson and Duke, 1979; Osmond et al., 1980; Nelson, 1984; Marek, 1988; Ceulemans and Saugier, 1991; Nobel, 1991; Dufrene and Saugier, 1993, in Larcher, 1995).

	μmol CO_2/m²/s
C4 plants	
Cenchrus ciliaris	68
Pennisetum typhoides	64
Hybrid sugarcane	64
Sudanese sorghum (*Sorghum*)	57
Maize	55
Herbaceous C3	
Comissonia claviformis	60
Triticum boeticum (a primitive wheat)	45
Typha latifolia	43
Rice (*Oryza sativa*)	40
Sunflower (*Helianthus annuus*)	28
Soybean (*Glycine max*)	27
Water hyacinth (*Eichhornia crassipes*)	20
Woody C3	
Willows (several species)	20-35
Poplars (several species)	20-30
Hevea brasiliensis	20-26
Oil palm (*Elais guineensis*)	20-25
Eucalyptus parviflora	15-20
Pinus sylvestris	17
Various plant groups	
Spring geophytes	15-20
Megaphorbiae	15-20
Plants of the undergrowth	2-10
Therophytes of the tropical deserts	20-40
Trees of shaded tropical forests	
leaves in sun	10-16
leaves in shade	5-7
Deciduous trees of cold temperate regions	
leaves in sun	10-15 (25)
leaves in shade	3-6

C4 plants nevertheless have some disadvantages. First of all, in limited light, that is light so limited that the rate of photosynthesis is much lower than the maximal rate and is in fact proportionate to the light intensity, C4 plants have quantitatively a smaller yield than C3 plants (Prioul, 1982).

This is explained not in terms of photochemical reactions strictly speaking, but in terms of the drift of energy towards phosphorylation. Throughout the Calvin cycle a C3 plant must consume 3 ATP to fix a CO_2 molecule in the equivalent form of CH_2O. A C4 plant must for the same

Table 2.2. Some data on water use efficiency and carbon use efficiency during
photosynthesis in C3, C4, and CAM plants (C. Black, in U. Luttge, 1993)

	C3	C4	CAM
Transpiration quotient (g H_2O/g carbon)	450-950	250-350	18-100 (for nocturnal fixation of CO_2) 150-600 (for daytime fixation of CO_2, which occurs for a short time)
Maximum net rate of photosynthesis (mg CO_2/m^2 leaf area/s)	0.41-1.10	1.1-2.2	0.027-0.360
Maximum growth rate (g dry matter/m^2 leaf area/d)	50-200	400-500	1.5-1.8

The transpiration quotient indicates the water mass (g) lost by transpiration when 1 g of
carbon is absorbed by photosynthesis.

Table 2.3. Effect of oxygen on growth of C3 and C4 plants (several works of
Bjorkman *et al.*, 1968, 1969, in Larcher, 1995)

	Growth of dry matter (mg/d/plant)	
	210 ml O_2/l	25 to 40 ml O_2/l
Maize (*Zea mays*, C4 plant)	127	147
Bean (*Phaseolus vulgaris*, C3 plant)	56	118

Measurements were taken at 320 µl/l CO_2, in medium light intensity (50-70 W/m^2) and at 24-
29°C.

result use more ATP, particularly for phosphorylate pyruvate: it requires
5ATP/CO_2. Also, when the global distribution of C3 and C4 plants is
examined, it is observed that C4 plants live in large numbers only in hot
places (Figs. 2.7 and 2.8). In cold or temperate cold countries C4 plants are
greatly outnumbered by C3 plants, which is explained by the relative
limitation of light in these plants and mostly by the effect of temperature
itself. In particular, it can be seen that C4 metabolism involves a great
many transfers of substances between cells and between organelles, and
many of the trans-membrane movements are dependent on the temperature. It will be seen later (see section 4) that if a plant cannot achieve a
good rate of photosynthesis at relatively low temperature, it will be
subjected to the phenomena of photoinhibition and photooxidation.

The high photosynthetic productivity of C4 plants in intense light is
explained by their facility in fixing CO_2, but also by the fact that in C3
plants the productivity is reduced by the phenomenon of photorespiration.

Finally, certain species may be intermediate between C3 and C4
(Angelov et al., 1993), as a function of climatic conditions. They may for
example switch from C3 to C4 metabolism in case of drought.

Table 2.4. Photosynthesis and light intensity in ambient CO_2, optimal temperature, and humidity (results of various authors)

Plant group	Compensation point of light I_c (μmol photons/m^2/s)	Saturation plateau I_s (μmol photons/m^2/s)
Terrestrial		
Desert plants		>1500
C4 plants	20-40	>1500
Cultivated C3 plants	20-40	1000-1500
Spring geophytes	10-20	300-1000
Sciaphyte phanerogams	3-10	50-200
Forest hemicryptophytes	5-20	300-600
Ombrophile tropical trees		
leaves in sun	15-25	600-1500
leaves in shade	5-10	200-300
young trees	2-5	50-150
Temperate deciduous trees		
leaves in sun	20-50	600-1200
leaves in shade	10-15	200-500
Conifers		
needles in sun	30-40	800-1100
needles in shade	2-10	150-200
Ferns in shade	1-5	50-150
Ferns in sun	50	400-600
Mosses	5-20	150-300
Lichens	50-150	300-600
Marine algae in cold or temperate environments		200-800
Some Antarctic Pheophyceae	2-5	100-150
Phaeodactylum raised in intense or low light	1-5	100
Some Arctic sub-glacial algae		4-20
Subaerial algae on old, dark walls	close to zero	10-20
Cavernicolous algae	close to zero	2-4

2.4.3. Photorespiration

For a long time it has been observed in many algae and higher plants that there is relative inhibition of photosynthesis by high levels of oxygen in the environment. This is called the Warburg effect. With just the normal level of O_2 in air, about 20 to 40% of the photosynthesis is inhibited, in relation to the activity in an environment containing only 1% rather than 21% of oxygen. The Warburg effect can be demonstrated in isolated chloroplasts, and it is thus directly linked to the mechanism of photosynthesis.

In fact, Rubisco fixes not only CO_2 but also oxygen (Fig. 2.9). This enzyme is a carboxylase as well as an oxygenase. The two substrates O_2 and CO_2 are in competition, and the product of the reaction is not the same. In the case of fixation of an oxygen molecule, a single PGA is produced, and a molecule of phosphorylate glycollate. After the loss of phosphorus, the glycollate leaves the chloroplast and passes into a

Grasses   Dicotyledons

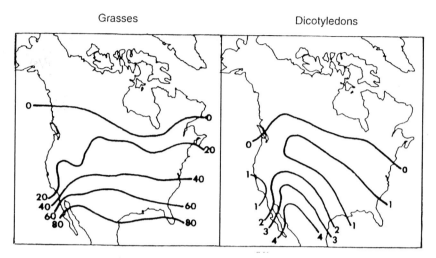

The figure at left represents the percentage of C4 among the monocotyledonous "grasses". The figure at right represents the percentage of C4 among the dicotyledonous herbs. There is a marked asymmetry between the coasts: in particular, the proportion of C4 grasses is low on the west coast of the United States because the summer there is relatively cool. There are few C4 dicotyledons, and they are seen to be higher in zones where the summer is more hot and dry.

Fig. 2.7. Geographic distribution of C3 and C4 plants in North America
(Larcher, 1995, according to numerous authors).

peroxisome, where it is oxidized into glyoxylate, which by transamination forms glycocol (Figs. 2.10 and 2.11).

Glycocol passes into a mitochondrion in which serine is formed, with the release of CO_2. The serine may yield another amino acid by transamination. After its return to the peroxisome, hydroxypyruvate forms, which will be subsequently reduced to glycerate in the chloroplast. Finally, a phosphorylation is produced that again gives a PGA that can return to the Calvin cycle (Fig. 2.12).

It must be noted that only one PGA "returns" for two phosphoglycollates that are formed by fixation of O_2 (in place of 2 PGA if it is CO_2 that is fixed). Thus, because of photorespiration, there is an overall loss of energy, in the form of ATP and at the same time because fewer sugars were formed. But amino acids are produced during the process of the close collaboration between three cellular organelles.

The advantages of photorespiration are discussed again in section 5. Although photorespiration may be high in C3 plants, it also exists in C4 plants but is not practical in the latter because the concentration of CO_2 in the sheath cells is very high, and it is therefore more effectively fixed than O_2 during the course of the competition. It must also be considered that O_2 is released mostly at the level of the mesophyll, which does not have Rubisco.

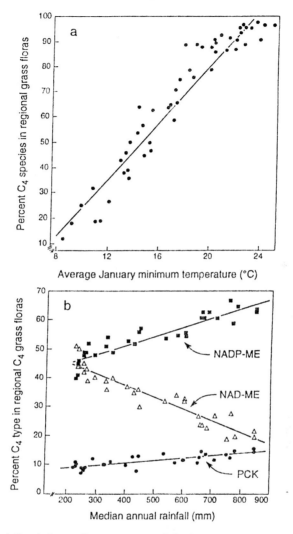

(a) A strong correlation between the percentage of C4 Poaceae and the heat in summer, according to the mean minimum temperature of January observed in the various regions of the continent. (b) The incidence of drought (annual rainfall) is marked by a very clear increase of the proportion of grasses with NAD malic enzyme (NAD-ME), but a reduction in that of grasses with NADP malic enzyme (NADP-ME). Few species have decarboxylation by phospho-enolpyruvate carboxykinase (PCK).

Fig. 2.8. The relationship between C4 metabolism and environment
in the Australian continent
(Hattersley, 1983, in Schulze and Caldwell, 1995).

Finally, after the formation of photosynthates by the Calvin cycle, we have the problem of transitory storage of energy (chloroplast amidon) and then its export from the leaves (Fig. 2.13).

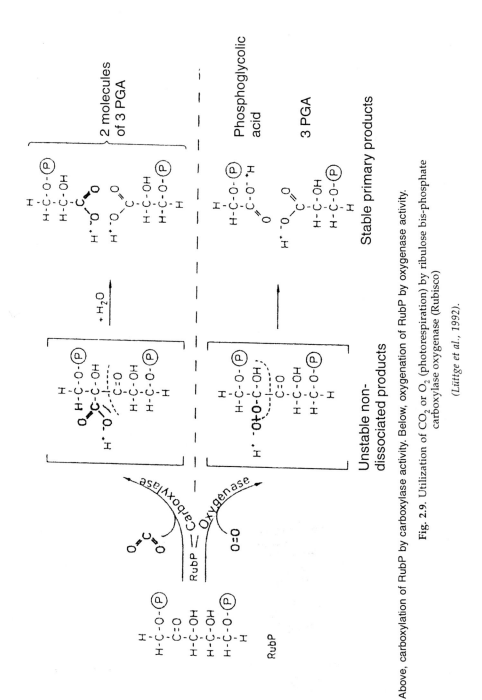

Fig. 2.9. Utilization of CO_2 or O_2 (photorespiration) by ribulose bis-phosphate carboxylase oxygenase (Rubisco)

Above, carboxylation of RubP by carboxylase activity. Below, oxygenation of RubP by oxygenase activity.

(Lüttge et al., 1992).

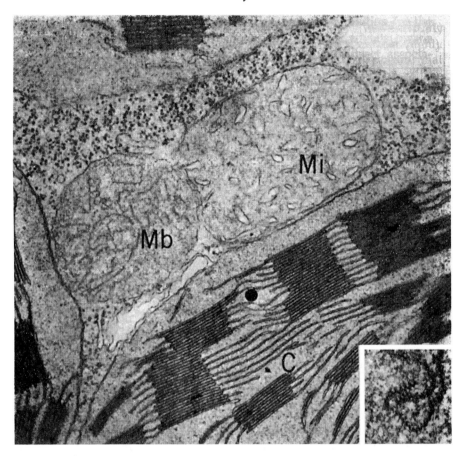

The photo shows an extra-fine section observed under transmission electron microscope. Part of a cell of tobacco leaf mesophyll is seen. The peroxisomes (Mb) are against the external envelope of the chloroplasts (C) and there are probably sites of passage of glycollate. The simple membrane of the peroxisomes can be observed, as well as their finely granular contents. As often in the higher plants, the mitochondria (Mi) are small and more or less spherical, and they have few crests. They are in immediate proximity to the peroxisomes and chloroplasts. We can also see the difference in size between ribosomes (of bacterial type) of the chloroplast stroma and the larger ones located in the cytoplasm.

Fig. 2.10. Close view of chloroplasts, peroxisomes and mitochondria
(Frederick and Newcomb, 1969).

2.5. Photosynthesis integrated with plant physiology: source-sink relationships

2.5.1. Role of amidon as transitory or long-term sink

During photosynthesis, the plant has an interest in automatically and quickly reinvesting its photosynthates in the construction of new leafy

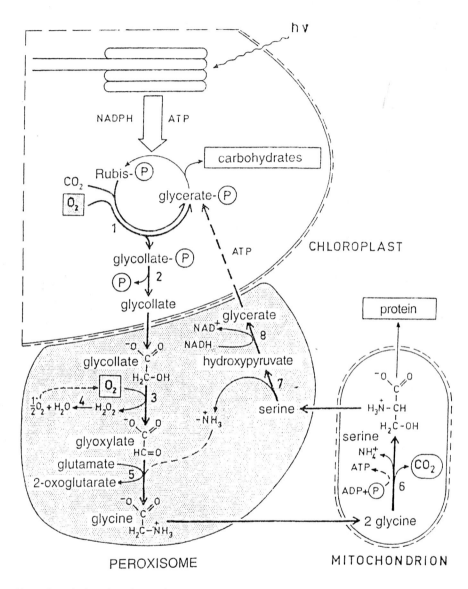

Here the glycine (or glycocol) comes from the transamination from glyoxylate in the peroxisome and passes into the mitochondria, where serine is produced. Serine into the peroxisome to form, by a new transamination, hydroxypyruvate and then PGA, which goes into the chloroplast. The malate oxaloacetate system allows exchanges of oxidoreducers between the mitochondria and the peroxisome.

Fig. 2.11. Role of peroxisome and mitochondria in photorespiration
(Amthor, 1995, in Schulze and Caldwell, 1995).

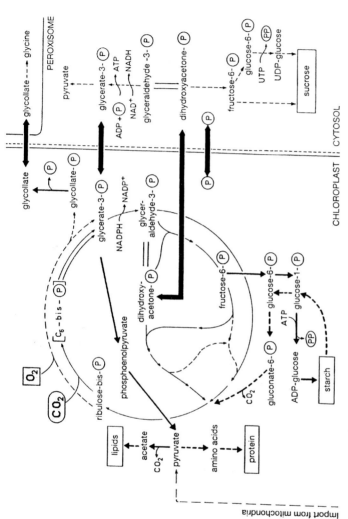

Fig. 2.12. The Calvin cycle and its communications with the extrachloroplast compartments
(Mohr and Schopfer, 1995).

The sugars are exported essentially in the form of dihydroxyacetone-phosphate (DHAP), which is partly exchanged with inorganic phosphate. Sugars of C4 to C7 cannot leave the chloroplast. Export of DHAP and PGA indirectly allows the maintenance of a good cytosolic concentration of ATP (Fig. 2.45). In this manner the chloroplast ultimately directly exports ATP. The exported glycollate is an element of photorespiration. The internal membrane of the chloroplast also has transporters of dicarboxylic acids and amino acids. The ADP-glucose, formed from glucose 6-P and glucose 1-P, contributes to the synthesis of amidon during the day. During the night and the evening, the amidon is hydrolysed into glucose 1-P, for the purpose of the cytosolic synthesis of saccharose, which will be transported to the sinks.

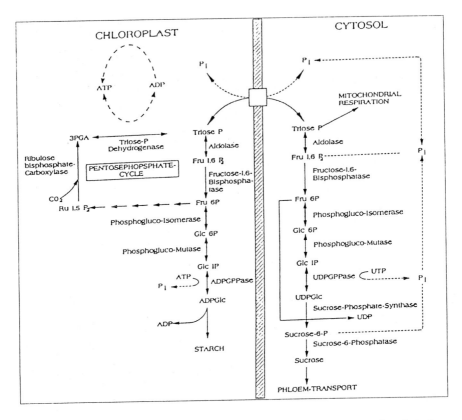

The triose-phosphates are exported mostly in the form of dihydroacetone phosphate in a cotransport that allows the entry of phosphate in the chloroplast. From cytosol, the sucrose that forms there subsequently passes into the apoplastic domain (phloem) in order to direct itself towards growing parts of the plant or for storage in the roots.

Fig. 2.13. Transitory storage of amidon in the chloroplast and the possibility
of its use for export
(*W. and E.D. Schulze, 1995*).

growth in order to increase its production. In fact, the problem is more complicated because the root system must also grow, and also the overall maintenance systems of the plant, which may involve the trunk and main branches of a tree as well as the formation of collenchyma or sclerenchyma in grasses. There are problems of transport and delay of synthesis for complex products such as lignin. Thus, short-term reserves are required for one to a few days.

There are periods during which photosynthesis is not possible or only a minimum basic metabolism is required. When the growing season returns, new leaves have to be produced. To meet all these requirements, short- or long-term reserves have to be built up, with systems of transport to pass the substances from source to sink or from one sink to another: for example, from a reserve organ to a growing organ.

The problem of storage, transport, and export of photosynthates is posed at the level of the chloroplast (Fig. 2.13), with competition between storage and export. Then, on the level of the whole plant, there is competition between storage and growth, thus between sinks.

If the plant is living under good conditions of nutrition—light, water, mineral salts—then competition may not be of much significance. But if the plant is living in poor light, what will it do? Reduce its growth?

There is a choice to be made that can be analysed by comparing wild plants and mutant plants, for example, that are incapable of synthesizing amidon. These mutants are also useful in the analysis of the ecophysiological advantages of synthesis and storage of amidon or other glucanes.

Several studies are available in this field (Schulze et al., 1991; Stitt, 1993; Stitt and Schulze, 1993) with the "plant drosophila", *Arabidopsis thaliana*.

2.5.1.1. Formation of amidon in the chloroplast
During the Calvin cycle, the chloroplast manufactures particularly C3 sugars and C6 sugars. The triose-phosphatases are exportable, although not effectively so, and they may continue in the Calvin cycle and thus again manufacture trioses. Or they may drift towards fructose-6-phosphate and then towards the synthesis of glucose-6-phosphate (outside the Calvin cycle) and continue by the formation of ADP glucose till the formation of amidon.

In the studies of Stitt and Schulze (1995), three mutants were used: (a) mutant of phosphoglucomutase (PGM), (b) mutant of a great part of the ADP glucose phosphorylase (ADPGPPase) activity, and (c) mutant resulting from a cross between wild and mutant ADPGPPase. We can see what happens during a daily turnover.

With wild *Arabidopsis* well supplied with nitrogen, the amidon concentration rose from 70 μmol/g fresh matter in the morning to 200 μmol/g fresh matter in the evening. If the nitrogen was low, the concentrations were higher but the daily accumulation was similar: from 200 to around 340 μmol/g fresh matter. If instead of nitrogen the light was reduced (from 600 to 80 μmol/m^2/s PFD), the accumulation fell from 130 to 50 μmol/g fresh matter.

In mutant "b", the ADPGPPase was low, and there was less amidon and less accumulation: 60 μmol/g fresh matter in the morning and around 90 in the evening. Mutant "a" did not have any amidon (because it could no longer make glucose 1-P) but from morning to evening there was a measurable accumulation of soluble sugars. It was also remarked that the accumulation depended on the age of the leaf: the young leaves of the rosette had more amidon than the old ones. The daily variations could not be exploited, but it may be considered that the young leaves also accumulate amidon from imported soluble sugars.

2.5.1.2. Amidon and growth

In strong light and with good nitrogen supply, a clear positive correlation is observed between the daily turnover of amidon and increase in biomass. Obviously, growth is stimulated when organic carbon is temporarily retained in the leaves, then exported and used during the night. If the nitrogen is low the effect is blurred, and it may be thought for example that the plant that has too much amidon will preferentially use it for producing roots (rather than to improve the growth of the whole plant), in order to be able to absorb more nitrogen in the future (to take the long view). This brings us to the role of amidon in the ratio of above-ground parts to underground parts.

A mutant without amidon that grows in a nitrogen-poor environment will produce an abundant root system (more than the wild) and more nitrogen will be available in its leaves than in the wild type. It seems clear that the increase of root growth allows better fixation of rare nitrogen, and apparently the presence of amidon does not have much to do directly with the stem/root ratio. It could be that resource allocation is not very precisely regulated.

2.5.2. General carbon equilibrium

A model may be attempted (W. and E.D. Schulze, 1995) in which the young leaves produce photosynthates to the same extent as adult leaves. Casper et al. (1986) showed that leaves of mutant PGM clearly respire more. Root respiration is nearly identical in the mutant and the wild plant.

In the mutant, there was only 57% of organic carbon invested in the growth of branches, because of excess respiration coming perhaps from the activation of alternative cyanoresistant respiration that may regulate the level of triose-P of hyaloplasm, liberate phosphate, and constitute a defence reaction.

In conclusion it can be said that amidon favours growth because it allows greater metabolic activity during the night and probably because it limits the respiration due to the low level of triose-P in cytosol. Regulation of the ratio of organic C to organic N does not seem to be closely linked to the presence of amidon.

2.5.3. Relationships between photosynthesis, storage and allocation

Arabidopsis represents a relatively simple case. The vegetative structure of the plant is mostly concentrated in a rosette (so that analysis of transport is simple), the life span is short, and a study can be based mostly on the analysis of daily movements.

It is useful to study other species, particularly agriculturally important herbaceous species, some of which, such as tobacco, have been modified in certain characters of their photosynthesis by genetic engineering.

When a character of photosynthesis is modified, a purer analysis is possible than when an environmental character is manipulated, which can never have an effect on just photosynthesis. For example, light also has morphogenetic effects.

Transgenic tobacco, which has been transformed with anti-sense sequences of rbcS (gene of the small subunit of Rubisco), can be used (Fichtner et al., 1995). By this means, we obtain mutants that have a fairly large quantity of the key enzyme of photosynthesis and thus a reduced capacity of photosynthesis, and few secondary effects. We can then study their reactions to various environmental constraints.

3. MAJOR FACTORS OF THE BIOTOPE ACTING ON PHOTOSYNTHESIS

3.1. Light

Light acts not only by its average value and by its spectral distribution (Fig. 2.14), but also by its fluctuations due to variable cloud cover or, in the undergrowth, scattered sunlight, or daily variations according to the photoperiod (Figs. 2.15 and 2.16). Part of the ultraviolet light may also be effective (Fig. 2.17).

An individual organism may receive very different levels of light on its above-ground parts. The upper crown of a beech tree has leaves in sunlight, and the lower part has leaves in shade (Tables 2.5 and 2.6). The same phenomenon applies to a maize plant in a field, with an alternating effect: the youngest leaves get sunlight, and the oldest (and also the smallest) leaves are in the shade, having been in sunlight when the plant was young.

In the case of a bush, light has an effect not only from above but also laterally. Whether it is in a meadow, a row of trees, or a small wood, the centre of the bush gets less light than the periphery.

Overall we can distinguish between plants in shade and plants in sunlight (Figs. 2.18 and 2.19). There are more or less extreme cases of shade-loving plants, and there are also intermediate cases (more difficult to define) of plants of semi-shade. Shade-loving plants (Fig. 2.20) are found in the undergrowth as well as in the lower stratum of a prairie formation, in shade at the base of cliffs, in the shelter beneath rocks, and other places. They are also found in the freshwater or marine aquatic environment at a depth depending on the clarity of the water (see Chapter 1, section 2). The low light is more or less modified in its spectral quality,

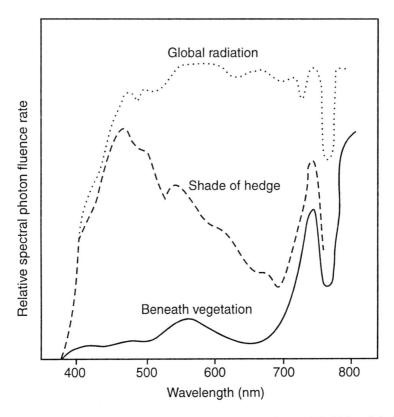

Dotted line, light on a prairie. Broken line, light at the edge of a wood. Solid line, light in the undergrowth. The wavelength (nm) is on the X-axis, and the photon flow per band passing from 1 nm is on the Y-axis (relative values).

Fig. 2.14. Spectral distribution of light in three environments
(Larcher, 1995).

which gives rise to pigmentary adaptations (see Chapter 1, section 2). We also have occasional variations in light, for example with dappled sunlight. Plants of semi-shade conditions that are found in a field under an isolated tree, in a small wood, and other places, can often show preferential illumination in one part of the day, the morning, for example, which brings about a chronobiology of the activity.

Sun-loving plants can overcome situations that are more or less extreme. Most of all, the insolation factor will greatly influence the temperature as well as the activity of environmental water (Fig. 2.21). Situations of stress could be numerous.

Energy loss in the form of heat increases greatly with the increase in incident light (Fig. 2.22). In strong sunlight, phenomena of photoinhibition could occur, involving particularly photosystem II (Fig. 2.23).

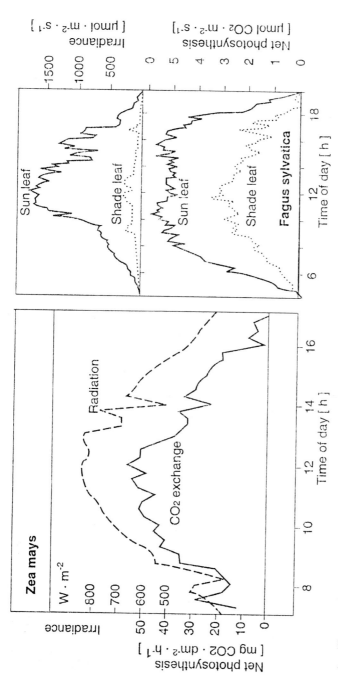

Fig. 2.15. Photosynthesis during the course of a day in a C4 plant and a C3 tree
(Hesketh and Baker, 1967; Schulze, 1970, in Larcher, 1995).

In maize (*Zea mays*), the net photosynthesis (Phn) closely follows each fluctuation of light, but still there is a slight departure from linearity from 500 W/m² (around 1200 μm/m/s PPFD). Most of all, a "fatigue" is observed in the plant at the end of the day. In the beech (*Fagus sylvatica*), there are leaves in shade and leaves in sunlight. The light, indicated in dotted lines, is 6 to 8 times as low in the leaves in shade. Activities in the shade (net photosynthesis) closely follow the fluctuations in light, which is always limited. For leaves in the sun, the photosynthesis curve does not follow the increase in light beyond around 1000 μm/m/s.

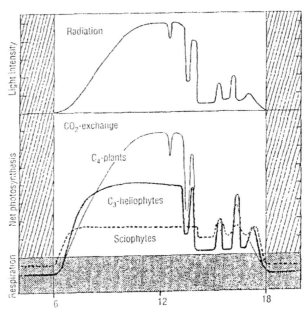

It is supposed that the morning is very clear, but that in the afternoon there are several cloudy interludes, which could reduce the illumination often by 5 to 20 times. Photosynthesis in C4 plants closely follows the fluctuations in light. In C3 heliophytes, photosynthesis remains best in the morning because they have better yield in limiting light than C4 plants and function better in low temperatures. In the middle of the day, however, the photosynthesis is saturated from 9 to 10 h. The C3 plant does not effectively use the light, which continues to increase, mainly because of its high photorespiration, and the limitation of its carboxylation rate. The sciaphytes photosynthesize better than the others when the light is low, that is, in the early morning and at the end of the day, and when the skies are cloudy. However, they use full sunlight much more inefficiently than the C4 plants or even the C3 heliophytes.

Fig. 2.16. Photosynthesis of several types of plants during the course of natural illumination in a day
(Larcher, 1995).

3.2. Photosynthesis and temperature

In terms of the overall photosynthetic process, there are photochemical reactions that are in principle independent of temperature and enzymatic reactions that are often closely dependent on temperature. In reality it is not so simple. Even in very low light, where the rate of photosynthesis depends only on rates of photochemical reactions, the influence of temperature is observed, however small it may be.

During electron transfer, there are stages of passage on liposoluble membrane substances, such as plastoquinones, the mobility of which between the cytochrome f+b6 complexes and the protein-quinone complexes that are electron donors may be considerably slowed down by temperature drops that diminish the membrane fluidity.

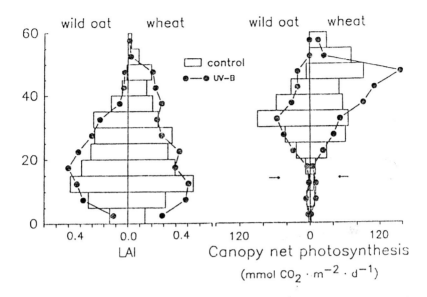

In a study of competition between wheat and wild oat, the influence of a UVB supplement was observed. During preliminary experiments it was observed that, in mixed culture, wheat was more competitive in the absence of UVB. The leaf length was observed to be less in the wild oat in the presence of UV. In monoculture, the behaviour and production of the two species are similar. On the other hand, it is notable that, after observations in the field and after modelling of photosynthetic production in mixed cultures, the leaf area index (LAI) improved in the upper parts of the wheat, contrary to what happened in oat, when UVB was added. The wheat plants were also clearly larger. In the mixed culture, the photosynthetic production modelled was always improved by UVB in wheat, and the contrary was true with wild oat.

Fig. 2.17. Photosynthesis in mixed cultivation, in presence of UVB
(Ryel et al., 1990, in Schulze and Caldwell, 1995).

Temperature may also have an impact on membrane permeability to all sorts of substances useful in photosynthesis, such as CO_2, inorganic phosphate, or malate. Temperature and light interfere in seasonal effects, particularly in woody plants with persistent leaves (Figs. 2.24 and 2.25).

The curves of photosynthesis as a function of the temperature, other factors being supposed to be at their optimum, show great differences between the species according to the living environment in which they have evolved. Also, it can be seen that the optimum (Table 2.7) is followed quite closely by inhibition of photosynthesis at high temperatures, a phenomenon that is explained by the general rule of intervention of thermal enzymatic denaturation (Fig. 2.26; see Chapter 1, section 4.2).

Often, photosystem II proves to be more sensitive to temperature than photosystem I (Fig. 2.23). Deterioration of photosynthesis may also be measured by the yield of fluorescence in low light (Fig. 2.27).

Table 2.5. Differences between leaves adapted to intense light and low light
(Larcher, 1995)

Characters	Leaves in sun	Leaves in shade
Leaf		
Area/dry weight	–	+
Thickness of palisade parenchyma	+	–
Thickness of lacunal parenchyma	+	–
Stomatal density	+	–
Chloroplasts/leaf area	+	–
Chloroplast		
Size	–	+
Stroma vol./thylakoid vol.	+	–
Thylakoids per grain	–	+
Chlorophyll/chloroplast	–	+
Chlor. a/chlor. b	+	–
P_{680}/chlorophyll	+	–
CF_0/CF_1/chlorophyll	+	–
P_{700}/chlorophyll	+ –	+ –
Lutein/zea + anthera + violaxanthins	–	+
Electron transport chain	+	–
Physiology		
PS I and PS II activities	+	–
Electron transport intensity	+	–
ATPase activity/chlorophyll	+	–
Quantum yield	+ –	+ –
Rubisco activity	+	–
Photosynthetic capacity	+	–
Mitochondrial respiration	+	–

Table 2.6. Characteristics of leaves in the shade and the sun, of beech and ivy, with
values of several traits showing the extent of the adaptation
(Lichtenthaler et al., 1981; Hoflacher and Bauer, 1982, in Larcher, 1995)

Character	Fagus sylvatica			Hedera helix		
	sun	shade	sun/shade	sun	shade	sun/shade
Leaf surface (cm²)	28.8	48.9	0.6			
Leaf thickness (μm)	185	93	2	409	221	1.85
Specific leaf area (dm²/g dry matter)				0.97	2.6	0.37
Stomatal density	214	144	1.5			
Stomatal conductance (cm/s)				0.65	0.33	2
Chlorophyll a/b	3.9	3.0	1.3	3.3	2.8	1.2
Net photosynthetic capacity (mg CO_2/dm²/h)	3.5	1.3	2.7	22.3	9.4	2.4
Light compensation point (W/m²)	2.5	1.0	2.5			
Light saturation (W/m²)	85	44	1.9			
Light saturation (μmol/m²/s)				600	250	2.4
Respiration in dark (mg CO_2/dm²/h)	0.5	0.16	3.1			

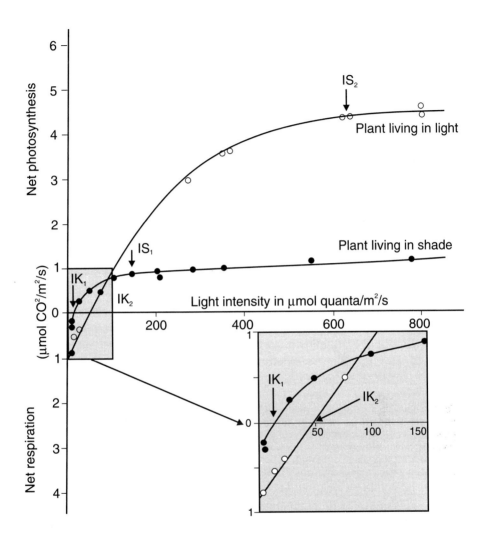

The study was done in a secondary tropical forest of Singapore. Although the light intensity saturation points were very different (IS$_1$ and IS$_2$), the curves of increase in photosynthesis with light when it remained limiting are similar. It is possible that in certain cases the yield of photosynthesis in limiting light is a little better in the shade plants (see the box). The possibility is still being studied. In any case, the net photosynthesis rates of species in the shade are clearly better in very limiting light, because as we have seen here the points of light compensation are much lower for the sciaphyte species (IK$_1$) than for the heliophyte species (IK$_2$), all other things being equal.

Fig. 2.18. Plants in shade and sunlight: major traits of their photosynthetic behaviour
(*Lüttge et al., 1992*).

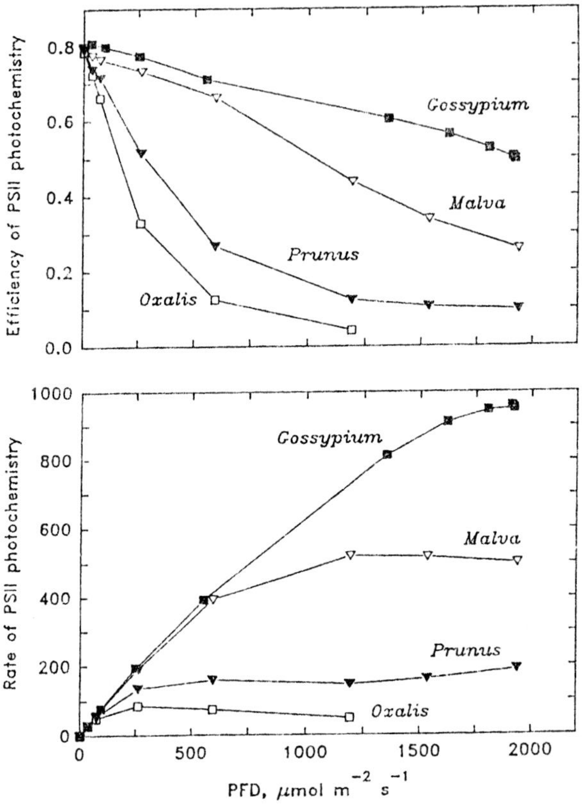

Top, an equivalent of quantum yield is estimated by analysis of kinetics of fluorescence (see section 4). Cotton (*Gossypium hirsutum*), which is the most heliophilous of the four species presented, still gives a good yield (3/4 of the optimum) at 1500 μmol/m²/s. The yield of *Oxalis oregana*, which is a vicariant of the European species *Oxalis acetosella*, becomes close to zero from 1000 μmol/m²/s onward. Bottom, the measurement of the rate of electron transfer from PS II shows a phenomenon of photoinhibition in *Oxalis*.

Fig. 2.19. Activity of photosystem II of heliophytes and sciaphytes
*(data from Bjorkman and Schafer, B. Demming-Adams and O. Bjorkman,
in Schulze and Caldwell, 1995).*

3.3. Photosynthesis and mineral nutrition

Apart from C, O, and H, nitrogen supply is essential for photosynthesis. As a general rule, the plant uses soil nitrate, and photosynthesis reduces it into NH_4^+, which is used in the synthesis of amino acids. The reducing activity of nitrates is very important and can even enter into competition, for energy utilization, with the synthesis of glucides.

The reduction of nitrates comprises three blocks of reactions (Fig. 2.28). First, nitrate is reduced to nitrite in the presence of nitrate reductase (NR) (Table 2.8), the synthesis of which is induced by NO_3^- and sup-

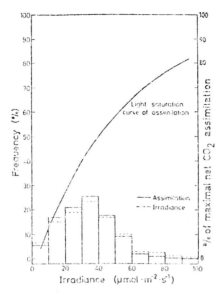

During a study of photosynthesis and transpiration in young beeches in the undergrowth, a particularly low light compensation point was observed: 2.8 µmol/m²/s PPFD. Here we see that the irradiation observed (n = 2.880) has a maximum frequency at 30-40 µmol/m²/s. The young beeches can achieve at least 50% of their maximal photosynthetic activity during more than half the days. Almost always there is a net photosynthesis. We also see according to the curve of photosynthesis as a function of illumination (light saturation curve of assimilation) that the young beeches only rarely have more than 80% of the potential activity that corresponds to illumination greater than 90 µmol/m²/s. There is likely to be only a little dappled sunlight; the old trees probably form a close, homogeneous canopy.

Fig. 2.20. Illumination in the undergrowth of an old beech grove
(*Stickan and Zhang, 1992*).

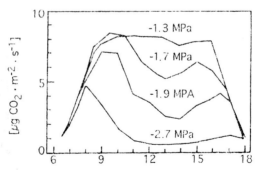

When the soil is well irrigated, soybean (*Glycine max*) photosynthesizes regularly throughout the day. The depression at noon appears when the soil becomes a little dry and leads to a drop in leaf water potential at midday. If the leaf water potential diminishes much more following the reduction of the soil water potential, photosynthesis is practically active only during the early morning hours.

Fig. 2.21. Daily photosynthesis and dryness of soil (*Rawson et al., 1978*).

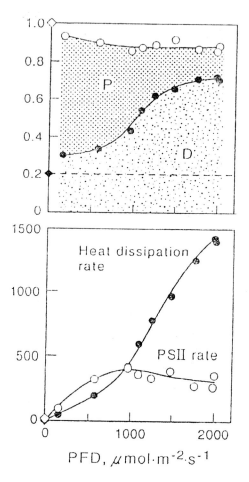

PFD, photon flux density. P, fraction of energy for photosynthesis. D, fraction of energy lost as heat. Values measured at midday on 10 *Euonymus kiautshovicus* leaves of different orientation.

Fig. 2.22. Energy loss in the form of heat compared to the energy converted by photosynthesis, as a function of incident energy
(B. Demming-Adams and W.W. Adams III, 1996).

pressed by NH_4^+. The reduction uses intermediate electron donors, particularly molybdenum (linked to pterine), FAD, and Cytb557.

$$NO_3^- + NAD(P)HH^+ \longrightarrow NO_2^- + NAD(P) + H_2O$$

Subsequently, nitrite reductase intervenes. It is a ferric heme enzyme and is rich in iron and sulphur. It is found in the stroma of the chloroplast

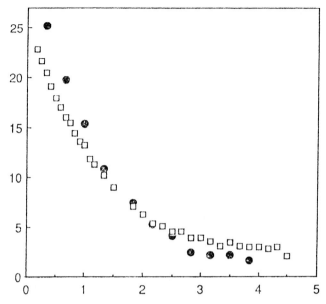

Measurements were taken on a population of willows (*Salix* spp.) growing "in the shade" in the laboratory (PPFD = 200 to 240 μmol/m²/s). Photoinhibition was triggered by a PPFD of 2000 μmol/m²/s at 25°C over 2 h. Activity recovered under moderate illumination (100 μmol/m²/s) interrupted only by short periods of dark red light (10 s) and dark (5 min.), which allowed fluorimetry and also measurement of the dark respiration. Photosynthetic activity was measured in limiting light (closed circles) under 100 μmol/m²/s, and it appeared closely correlated in its variations, with Fv/Fm (open squares). Fv/Fm is thus a convenient and rapid measurement of a state of photoinhibition. X-axis: hours. Y-axis: arbitrary units.

Fig. 2.23. Photoinhibition and measurement of fluorescence to estimate the
recovery of activity
(*E. Ogren, 1991*).

and initiates the following reaction, by means of the presence of ferredoxin (Fd):

$$NO_2^- + 6FdRed + 8H^+ \longrightarrow NH_4^+ + 6Fdox + 2\ H_2O$$

Like NO_2^-, NH_4^+ does not remain in the free state in high concentration. It gives rise to a group of reactions in the chloroplast comprising the GS/GOGAT cycle. The reaction also involves Fd and can be expressed overall as follows:

$$NH_4^+ + 2\ oxoglutarates + ATP + 2\ (H) \longrightarrow H_2O + H^+ + glutamate + ADP + Pi$$

The transaminations, which form the other amino acids, are mainly

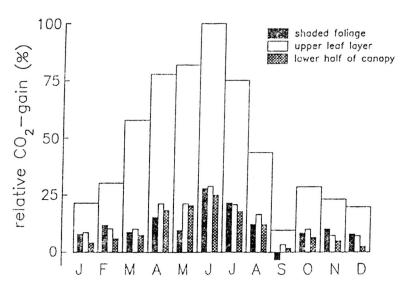

The study was conducted to test a system of modelling the production of a forest cover. Here, the kermes oak, *Quercus coccifera*, and a leaf area index of 5.4 were used. The foliage was divided into three parts: leaves in the sun (13% of the total leaf area), leaves in the shade, and intermediate leaves. Production was estimated only for clear days. The large white rectangles represent the overall assimilation of CO_2, in percentage relative to the best month. The small rectangles represent the contribution of each part: leaves in sunlight (white), intermediate leaves (grey), and leaves in shade (black). It can be seen that the leaves in the shade may be more productive than leaves in the sun, for several winter months, perhaps because phenomena of photoinhibition occur in the upper parts of the foliage.

Fig. 2.24. Seasonal variations of photosynthesis for different parts of
leaves of an evergreen tree
(Caldwell et al., 1986, in Schulze and Caldwell, 1995).

extrachloroplastic. NH_4^+ can be assimilated by the roots into glutamine. NH_4^+ is always in very low concentrations in the chloroplasts because it acts as a decoupler of photophosphorylation.

The reduction of sulphate, which is the form of sulphur most often present for plants, occurs in the presence of ATP.

$$SO_4^- + 8e^- + 9H^+ \longrightarrow (HS^-) + 4\ H_2O$$
$$ATP$$

SO_4^{--} is activated by ATP in the form of adenosine-phosphosulphate, which is fixed on a carrier protein. Sulphate reductase (Fig. 2.29) may also intervene and, through reduced ferredoxin, form the group SH, which is transferred to O-acetylserine with the formation of cystein.

Phosphate is, of course, very important in photosynthesis (Fig. 2.30). It is generally in very low concentration in soil water, and the presence of mycorrhizae helps greatly in the phosphate nutrition of trees. Phosphate

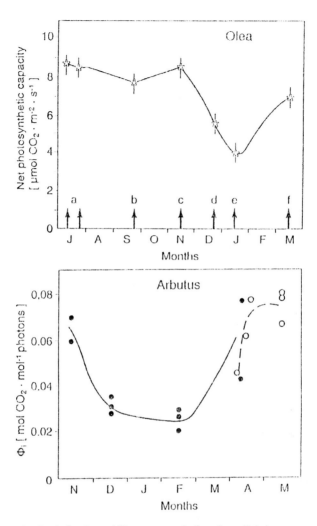

Left, the photosynthesis of olive trees (*Olea europaea*) of northern Italy is measured under 600 μmol/m²/s PPFD (Larcher, 1961). The letter a (July) shows data including the leaves that have developed this year, b (September) the time after a short period of summer drought, c (November) a very rainy period, d (December) a time before the first frost, e (January) very cold nights (temperatures as low as −5°C) and high content of soluble sugars in the leaves, and f (March) photosynthesis resuming a high rate but still with leaves of the preceding year. Right, the quantum yield of arbutus (*Arbutus unedo*) in Portugal (Beyschlag et al., 1990). P_1 is calculated by the slope at the origin of the photosynthesis/light curve, in non-limiting CO_2. The closed circles represent the old leaves, and the open circles represent the new leaves that appear from April onward.

Fig. 2.25. Photosynthesis during the season in Mediterranean trees
having persistent leaves
(Larcher, 1995).

Table 2.7. Net photosynthesis in saturated light and under ambient CO_2 as a function of temperature (Larcher, 1995)

Higher taxonomic group	Lower limit of Pn > 0, °C	Optimum of Pn	Limit of Pn > 0
Herbaceous angiosperms			
C4 plants	5 to 10	30 to 40	50 to 60
Cultivated C3 plants	–2 to 0	20 to 30, sometimes 40	40 to 50
Heliophytes	–2 to 0	20 to 30	40 to 50
Sciaphytes	–2 to 0	10 to 20	to 40
Desert plants	–5 to 5	20 to 35, sometimes 45	45 to 50, sometimes 60
During the day	–2 to 0	sometimes 20, 30 to 40	45 to 50
During the night	–2 to 0	10 to 15 (23)	25 to 30
Geophytes, spring annuals	–5 to –2	10 to 20	30 to 40
Plants of high mountains	–6 to –2	15 to 20	38 to 42
Aquatic plants	0	often 20 to 30	45 to 52
Woody spermatophytes			
Evergreens with large leaves, subtropical	0 to 5	25 to 30	45 to 50
Sclerophylls of dry climate	–5 to –1	20 to 35	42 to 45
Deciduous trees of temperate zone	–3 to –1	20 to 25	40 to 45
Conifers with persistent needles	–6 to –3	10 to 25	35 to 42
Shrubs of the tundra	to –3	15 to 25	40 to 45
Mosses			
Temperate regions	to –5	10 to 20	30 to 40
Arctic regions	to –8	5 to 12	to 30
Lichens			
Cold regions	–10 to –15	8 to 15	25 to 30
Deserts	to –10	18 to 20	38 to 45
Tropics	–2 to 0	to 20	25 to 35
Algae			
Algae of snow	to –5	0 to 10	30
Thermophilous algae	20 to 30	45 to 55	65

can, in a favourable season, be stored by the plant in the form of phytine, which is phytate of calcium and/or of magnesium. K^+, Fe^{++}, and Zn^{++} can also be fixed on phytic acid, which is incorporated within the proteic bodies (grains of aleurone) in the cells.

$$\text{myo-inositol} \longrightarrow \text{phytic acid}$$
$$PO_4^-$$

Iron, calcium, and magnesium are relatively abundant materials but at various antagonist degrees. Acid environments show the presence of mobile iron that can be assimilated, but very little calcium and

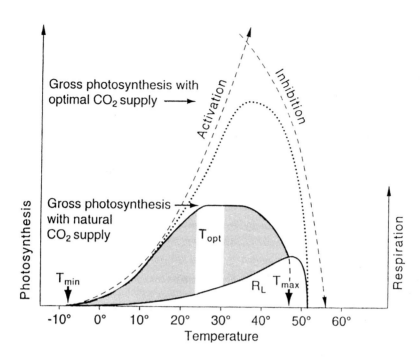

Gross photosynthesis with optimal CO_2 supply ⟶

Activation

Inhibition

Gross photosynthesis with natural CO_2 supply

T_{opt}

T_{min}

R_L T_{max}

Photosynthesis

Respiration

-10°   0°   10°   20°   30°   40°   50°   60°
Temperature

The conditions are non-limiting light and low or high CO_2. In low CO_2, the net photosynthesis (shaded part of graph) falls with temperature, because of the increase in light respiration (R_L). In high CO_2 (dots), from 0 to around 30°C, photosynthesis increases rapidly, with a Q10 that may reach around 2.5 and is often higher toward 20°C than around 10°C. Photosynthesis reaches a maximum, then falls because of the significance of mitochondrial respiration and especially photorespiration. Moreover, enzymatic activities are disturbed. Enzymatic and membrane denaturation intervene from around 40°C. T_{min} and T_{max} are the cold and heat limits of net photosynthesis. All these indications vary a great deal according to species and ecosystems.

Fig. 2.26. Variation of photosynthetic productivity as a function of the temperature
(Larcher, 1995).

magnesium. On the other hand, particularly for calcifugous plants, a rather neutral medium could be rich in Ca and Mg, but the assimilation of iron by certain plants would be problematic. Calcium as well as Cl^- influences the stability and functions of proteins of photosystem II on the electron donor side.

Manganese and copper intervene in the electron transport in photosynthesis: Mn for the electron donor side of photosystem II, and Cu as an active group of plastocyanin.

Boron deficiency often poses problems in agriculture, for example in viticulture. Boron seems linked indirectly to photosynthesis in the form BO_3^{---}, which is an inhibitor of the oxidation cycle of phosphate pentoses.

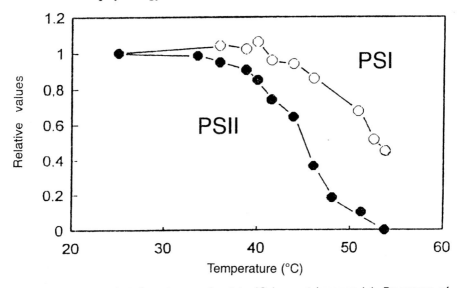

The study was conducted on leaves of potato (*Solanum tuberosum* L.). By means of fluorimetry (see section 4), the maximum quantum yield (closed circles) can be calculated from the PS II photochemical reaction. The PS I activity (open circles) was measured (change in absorption at 820 nm, ΔS/S, before the stress: 15×10^{-3}) by oxidation of P_{700} by a flash of dark red. The measurements were taken 15 min. after thermal treatment. The values are relative, 1 representing the untreated samples.

Fig. 2.27. Behaviour of photosystems under thermal stress.
(*Havaux, 1996*).

3.4. Photosynthesis and pH

The pH of symplastic fractions has, of course, a direct impact on photosynthesis, but so does the pH of the apoplast. The chloroplastic stroma is particularly sensitive to pH, and this pH is around 8 in the light and 7.4 in the dark. It is therefore very close to that of cytosol. These variations may seem small but, for example, the activity of fructose bisphosphatase (FBPase) is optimal at a pH of 8 and close to zero below a pH of 7.5 (Woodrow et al., 1984).

The pH of the apoplast changes throughout the life of the plant cell—during its development (particularly during cell enlargement) and during aggressions from the medium, such as pollution by sulphurous gas. There is an external optimal pH zone that seems quite extensive if one works with protoplasts (Pfanz, 1987) but lesser with tissues in which the apoplast has been only slightly disturbed. The effects of external (H^+) must depend on the permeability of the plasmalemma. Certain algae, such as *Dunaliella acidophila*, have optimal photosynthesis at a pH of 1 (Gimmler et al., 1989). Most species of algae are much less acidophilic (Fig. 2.31).

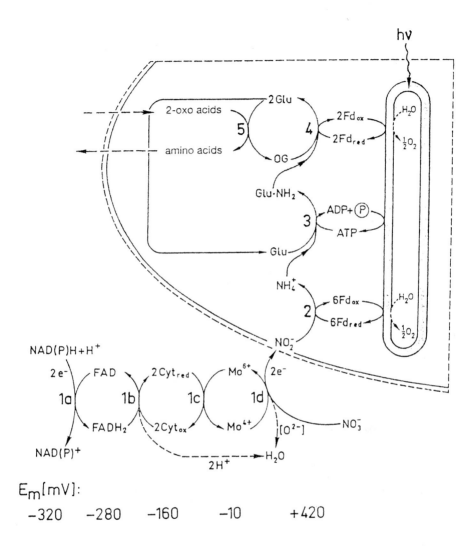

Fig. 2.28. Reduction of nitrate
(Lea and Miflin, 1974, in Mohr and Schopfer, 1995).

The reduction of nitrate into nitrite occurs in cytosol, by means of nitrite reductase, located at the end of a redox chain. The E_{m_7} of the chain is indicated. That of nitrite reductase was measured in *Chlorella*. In the chloroplast the reduction of nitrite by nitrite reductase involves the oxidation of ferredoxin (Fd) to reduce NO_2^- into NH_4^+. The ammonium is linked by glutamine synthetase (GS) (reaction 3) with glutamate to form glutamine (glu-NH_2) with consumption of ATP. Glutamine reacts with 2-oxoglutarate (OG) via glutamate synthetase (GOGAT) to yield **two glutamates**, with consumption of electrons from ferredoxin. The chloroplast (reaction 5), with 2-oxoacids, may realize transaminations.

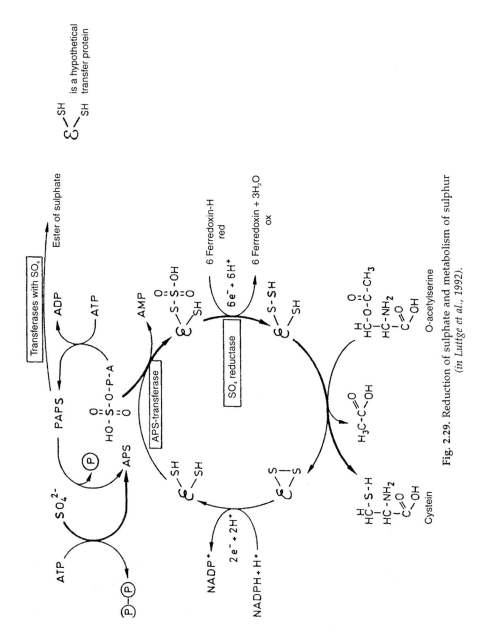

Fig. 2.29. Reduction of sulphate and metabolism of sulphur
(*in Luttge et al., 1992*).

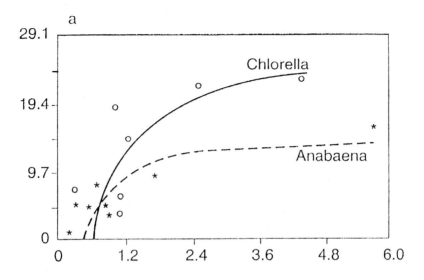

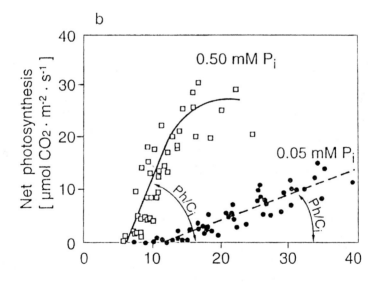

Top, the effect of the intracellular concentration in P_i on the growth (X-axis, relative units), in a unicellular alga and in a nitrogen-fixing cyanobacterium (Senft, 1978). Bottom, the relation between photosynthesis and internal pressure of CO_2. It can be observed that the efficiency of carboxylation (Ph/Ci) is mediocre in conditions of phosphorus deficiency (Laver et al., 1989).

Fig. 2.30. Photosynthesis and mineral phosphate uptake
(Schulze and Caldwell, 1995).

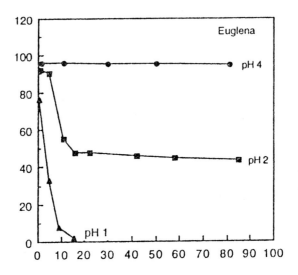

Euglena tolerate acidic water, but within limits. Although the optimum zone for stable production is pH 4, at pH 2 Euglena remains stable, but with reduced photosynthesis. A medium that is too acid imposes additional energy expenditure. X-axis: min.

Fig. 2.31. Photosynthesis of a preferentially acidophile alga as a function of pH
(Pfanz, 1995, in Schulze and Caldwell, 1995).

4. INPUTS OF FLUORIMETRY

4.1. Analysis of rapid kinetics of fluorescence in modulated light

4.1.1. Introduction

It is mostly the methods of measurement and analysis of chlorophyll fluorescence *in vivo* that have been developed in recent years. Much progress has also been made in photoacoustic measurements, which are not discussed here. Thus, new methods of analysis of chlorophyll fluorescence that are highly useful in ecophysiology have been perfected (Schreiber et al., 1995).

Granted that variations of thermal losses are gradual, the rapid kinetics of induction of fluorescence intervening during an instantaneous passage from dark to light can be interpreted by changes of photochemical quenching, which is determined by the concentration of open reaction centres, that is, centres that can liberate an electron at the given instant.

With all the centres open, all the primary electron acceptors, Q_A, are oxidized. If the electron flux is inhibited, all the Q_A are reduced, and all the centres are closed. Measurement of fluorescence must therefore include an estimation of minimum fluorescence, Fo, which is due exclusively to the fluorescence of antennae. The centres do not fluoresce, all of them being open.

100% $Q_A Ox \rightarrow$ Fo Fm – Fo = Fv = variable fluorescence

100% $Q_A Red \rightarrow$ Fm = maximum fluorescence observed, e.g., in presence of DCMU Fm/Fo may reach 5 to 6, i.e., Fv/Fm = 0.8 to 0.83.

The rate of reduction of Q_A is greater than that of oxidation of PQ (acceptor with 2 electrons) and a pool of acceptors may be represented by the area of increased fluorescence (Fig. 2.32). The DCMU is a block between Q_A and Q_B. Thus, the pool of the acceptor with one electron, i.e., the pool of Q_A, is measured.

There is cooperation between all the centres, except the PS IIB centres (dispersed in the intergranular lamellae).

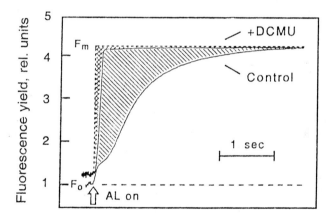

After a long period of darkness, all the pools of electron acceptors are oxidized. In light (actinic light, 650 nm, 100 $\mu M/m^2/s$), we "instantaneously" reach the Fo level, and then the fluorescence increases according to a kinetic complex (control) following a roughly sigmoid shape up to the Fm level, where all the PS II reaction centres are closed and where all the electron acceptor pools are reduced. The hatched area to the left is thus proportionate to the size of electron acceptor pools. If the experiment is repeated in the presence of DCMU 10^{-5} M, which inhibits the passage of electrons between Q_A and Q_B, the Fm level is reached much more quickly. The centres close very quickly because they cannot evacuate the electrons beyond Q_A, which is a very small electron pool. If we admit that the number of Q_A is equal to the number of centres (D1 protein), the hatched area to the left (by far the smallest) is proportionate to the number of PS II centres.

Fig. 2.32. Pools of electron acceptors of PS II and induction of fluorescence
(Schreiber et al., 1995, in Schulze and Caldwell, 1995).

If the light is not very intense, Q_A^- can reach equilibrium more or less with Q_B^- before Q_B^{--} can be formed, and we thus have a small transitory peak (old level I). I is different if the light is intense without reaching Fm (which does not change). First there is a level I_1 corresponding fairly well to Q_A^- in "equilibrium" with Q_B^-, then a level I_2 that later corresponds to the arrival of electrons on the donor side ("thermal" phenomenon) (Fig. 2.33). A saturating light of around 200 ms is required to reach Fm level, in the higher plants.

The slow phase of fluorescence (Fig. 2.34) indicates the effect of ascorbate peroxidase, and of course the formation of ATP for the activity of the Calvin cycle.

From the beginning of illumination, the O_2 emitted by PS, II can form O_2^-, a superoxide ion that will be decomposed by the superoxide dismutase (SOD) and APOx. The ascorbate peroxidase may also intervene, which results in slow oscillations of O_2 emission and pH gradient.

The pH gradient, the xanthophyll cycle, the membrane conformation, and chlorophyll-protein complexes provide the main part of non-photochemical quenching.

In the stationary state, the low fluorescence comes from two quenchings:

- photochemical Q, which comes from electron flux of the PS, and
- non-photochemical Q, coming from pH gradient, biochemical reactions, cycle of xanthophyll, etc., which yield heat.

4.1.2. Practical use based on "flashes" of intense light

From the preceding data, the saturation pulse method was developed. If there is intense light, Q_A is 100% reduced and the photochemical quenching equals zero. Only the non-photochemical quenching remains.

It is necessary to measure Fo and thus have a non-actinic light to keep Q_A totally oxidized, and also to have Q_A 100% reduced, even if the plant has a very good electron transfer. The pulse amplitude modulation (PAM) fluorimeter was developed for this purpose.

With samples adapted to dark, we obtain Fo and Fm.

(Fm – Fo)/Fm = Fv/Fm is the measurement of potential maximum quantum yield of PS II.

If a sample is illuminated long enough with pulses (very short periods of intense light), we get Fm', which can also be written as F'm, which is the maximum fluorescence in specific conditions (Fig. 2.35).

Fm – Fm' represents the non-photochemical quenching = qN. Fm' – F= photochemical quenching = qP, F representing the stationary basic fluorescence.

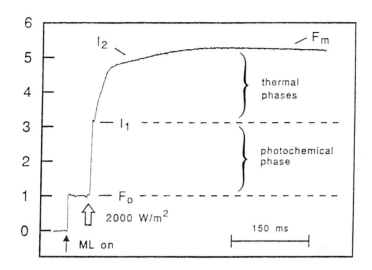

Fig. 2.33. Rise in fluorescence up to the Fm level
(Schreiber et al., 1995, in Schulze and Caldwell, 1995).

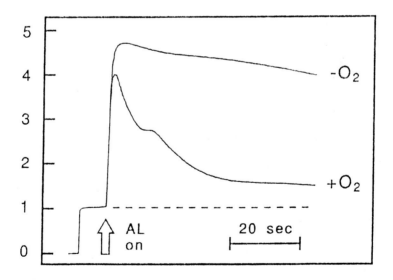

Fig. 2.34. Slow phase of fluorescence as a function of the presence of oxygen
(Schreiber et al., 1995, in Schulze and Caldwell, 1995).

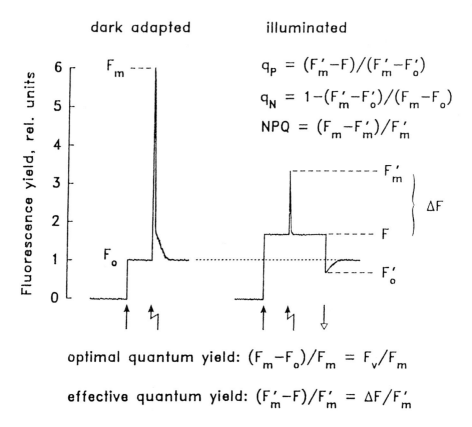

The most common formulae are indicated. See text for definitions and use of formulae.

Fig. 2.35. Standard nomenclature of different parameters of fluorescence level, obtained in low light and actinic light
(Schreiber et al., 1995, in Schulze and Caldwell, 1995).

4.1.3. *Energy yield of PS II and non-photochemical quenching*

Fm′ − F/Fm′ = 1 − F/Fm′ (if all the centres are open, F = Fo′) and (Fm′ − Fo′)/Fm′ = Fv′/Fm′, which is equivalent to an effective energy yield.

If the non-radiative dissipation (NRD) = 0, the intrinsic yield is Fv/Fm. The quenching of Fm into Fm′ caused by NRD is called non-photochemical quenching (NPQ).

The percentage of closed centres is Qr/Qt.

NPQ = (Fm − Fm′)/Fm′ = Fm/Fm′ − 1; Qr/Qt = 1 − Qox/Qtot = (F − Fo′)/(Fm′ − Fo′)

If the only cause of the quenching of Fm is the increase in NRD, then NPQ calculated in this manner will be proportionate to NRD (and also to the "quenching agent", J). NPQ is mostly linked linearly to the excess light intensity, in a wide range.

Since Fo is influenced by ΔpH, and thus by qN, it is necessary to measure Fo' to get qN and qP. Fo' can be determined by putting the organism in dark and then in low, dark red light that will oxidize Q_A by PS I. Calculation of Fo' is indispensable to finding out the extent of the "opening" of PS II by calculation of qP.

On the other hand, the non-radiative dissipation can also be calculated without Fo':

NPQ = (Fm – Fm')/Fm'

It must be noted that this expression of NPQ is based on the matrix model of organization of the antenna and admits the existence of "NPQ traps", e.g., zeaxanthin (Figs. 2.36 and 2.37).

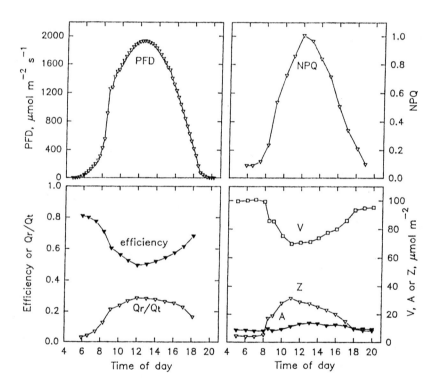

PFD represents the variation of illumination in California, NPQ is non-photochemical quenching, Q_R/Q_T is the degree of closure of PS II centres. The curves A, V, and Z are for antheraxanthin, violaxanthin, and zeaxanthin.

Fig. 2.36. Correlations between light, state of PS II, non-photochemical quenching of fluorescence, and the pool of xanthophylls
(according to the original results of B. Demming-Adams, O. Bjorkman, W. Bilger,
W.W. Adams III, S.S. Thayer and C. Shih, in Schulze and Caldwell, 1995).

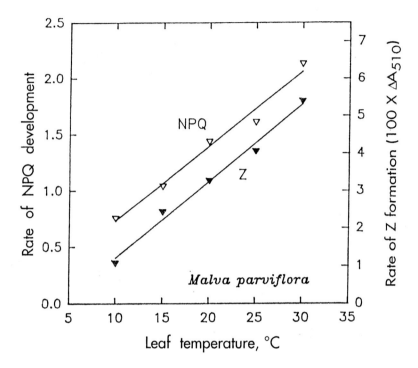

The leaves of *Malva parviflora* are pretreated in dark, then subjected to an actinic light of 2000 μmol photons/m²/s. The increases in NPQ and Z (zeaxanthin) are measured at the end of 1.5 min. of illumination. Z is measured after Δabs at 510 nm.

Fig. 2.37. Effect of temperature in non-photochemical quenching and level of zeaxanthin *(Bilger and Bjorkman, 1991, in Schulze and Caldwell, 1995).*

Table 2.8. Average nitrate reductase activity (μmol NO_2^-/g dry matter/h) of different types of plants, or plants of particular habitats (Gebauer et al., 1988; Stadler and Gebauer, 1992; Downs et al., 1993, in Larcher, 1995)

Plants or habitats	Leaves	Underground parts	Reproductive organs
Ruderal plants	14	1	4.5
Nitrogen-rich soil	7 to 13	0.6 to 1	1.6 to 4
Riparian grasses	9	0.6	1.6
Prairies	3 to 4	0.3 to 0.5	0.5 to 0.7
Prairies and alpine herbs	1	0.3	0.9
Lawn on poor chalky soils	0.8	0.1	0.2
Lawn on poor siliceous soils	0.2	0.1	0.3
Heathland	0.06 to 0.1	0.04 to 0.07	0.8
Pine seedlings	0.4	1.6	
Acer rubrum seedlings	1.5	0.7	
Ash trees aged 10 to 15 years	1.4	0.1	

4.1.4. Determination of quantum yield and rate of PS

The empirical equation of Weiss and Berry (1987) is based solely on fluorescence and is closely parallel to the measurement of CO_2.

$$\text{Rate of PS} = J = I \cdot qP \,(m - b \cdot qN)$$

I is the photon flux, m and b are empirical constants specific to each plant.

$$\text{Quantum yield of PS II} = \varnothing_{II} = J/I = qP\,(m - b \cdot qN) = qP \cdot \varnothing p$$

Thus, m - b·qN corresponds to the quantum yield (Rdqt) of open PS II centres: $\varnothing p$.

$$\varnothing p = \varnothing_{II}/qP = m - b \cdot qN = \varnothing p$$

m = maximum Rdqt of the sample adapted to darkness (qN = 0) = $\varnothing_{Po}$, and m − b represents the minimum Rdqt (qN = 1). It is currently believed that m - b is negligible. Since m = $\varnothing_{Po}$ (because qN = 0), therefore $\varnothing p = \varnothing_{Po}(1 - qN)$.

$$\varnothing p = m - m \cdot qN = \varnothing_{Po} - \varnothing_{Po}\, qN = \varnothing_{Po}(1 - qN) = \varnothing p$$

$\varnothing_{Po}$ is obtained when all the centres are open and ranges from 0.3 to 0.35 (Weiss and Berry, 1987).

$$\varnothing p = \varnothing_{Po}(1 - qN)$$

A value of 0.35 corresponds to a potential quantum yield of PS II, as Fv/Fm = 0.833 at the optimum. The correspondence with 0.35 is possible because we have two photoreactions so a 0.5 factor and around 84% of the incident light is absorbed by a leaf, so a 0.84; factor 0.833 = 0.35(1/0.84 × 1/0.5).

Since 1 − qN = (Fm′ − Fo′)/(Fm − Fo) = Fv′/Fv, therefore $\varnothing p = \varnothing_{Po} \cdot (Fv'/Fv)$.

This gives a drop in Fv′ that is parallel to the drop in Rdqt. The inactivation of PS II centres is parallel to a non-radiative process of dissipation. Apparently, the drop in Rdqt, $\varnothing_{Po} - \varnothing p$, is proportionate to qN.

Krause and Weiss (1991) proposed that acidification (pH < 5.5) of thylakoids led to inactivation of PS II centres and the centres were not fluorescent by a powerful phenomenon of non-radiative dissipation. The loss of Ca^{++} may be implied (Krieger, 1992). When the centres lose their Ca^{++}, Q_A^- and P_{680}^+ recombine, because of which everything is transformed into heat.

In parallel, a weakness on the donor side is noted, and there is an increase in luminescence. And, of course, Fm decreases and qN increases.

It can also be demonstrated that $\varnothing_{II}$ is well represented by the following equation (Genty et al., 1989):

$$\emptyset_{II} = qP \cdot (Fv'/Fm') = (Fm' - F)/Fm' = \Delta F/Fm' = Fx'/Fm'$$

thus $\emptyset p = \emptyset_{II}/qP = Fv'/Fm' =$ "actual" quantum yield of PS II, effective at the moment studied. The advantage of $\emptyset_{II} = qP \cdot (Fv'/Fm')$ is that there is no need to measure Fo'.

The Rdqt is represented (determined) by the product of the "opening" of "PS II centres", i.e., qP multiplied by the reactivity or efficiency of open centres, represented by Fv'/Fm', this second factor involving a more or less great non-radiative dissipation at the level of antennae.

Genty et al. (1989) observed a linear relation between Rdqt of CO_2 fixation and $\Delta F/Fm'$, $\Delta F = Fm' - F = Fv'$, so Fv'/Fm', in conditions of saturating CO_2 and with 1% O_2.

But if there is 21% O_2, the non-linearity becomes considerable and not only the fixation of CO_2, but also photorespiration and other influences of O_2 intervene in $\Delta F/Fm'$.

A fairly regular curvilinearity can be observed between $\Delta F/Fm'$ and Rdqt of the O_2 emitted, but with 5% of the CO_2. That is, the oxidase activity of RUBPcase is nearly null and the residual non-linearity comes from a Mehler reaction (reduction of O_2 and then of H_2O_2).

4.1.5. Fluorescence as indicator of electron flux not integrated in photo synthesis

Fluorescence also reacts at the level of internal CO_2 and O_2 of the cell, which becomes very important, above all, if the stomata are closed because there is very little CO_2 (Fig. 2.38). These electron flows can help prevent photoinhibition. Photoinhibition acts mainly by oxidation and breaking down of essential compounds such as protein D1 (Trebst, 1995), attacked by single oxygen molecules.

Two aspects can be distinguished by means of fluorescence:

• PS II electron flux on the acceptor side and
• the membrane pH gradient.

If there is CO_2, $\Delta F/Fm'$ again increases quite rapidly. The initial drop depends on the presence of O_2 and reflects the formation of a pH gradient. The ultimate increase in Fm' comes from the starting up of the Calvin cycle and the increase in O_2 concentration.

If CO_2 is absent, the Calvin cycle cannot run and the organism soon reaches a nearly stationary state with low fluorescence that is characterized by an enormous non-photochemical quenching and nevertheless a significant electron flux.

From this manipulation it can be seen that some phenomena are not detected by infra-red gas analysis (IRGA) or the O_2 electrode, but they are seen by analysis of quenching of fluorescence.

In particular, the coupling between the Mehler reaction and the ascorbate peroxidase (APO) activity manifested by nearly null gaseous

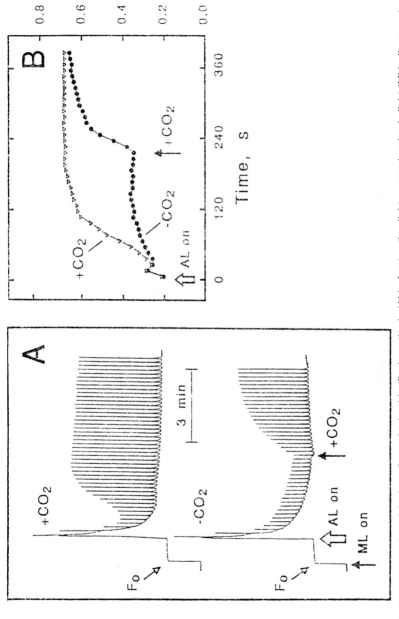

Part A represents the original readings obtained by flash methods (AL) of saturating light, a very low basic light (ML) allows us to define the Fo level. From the successive peaks we can calculate the successive values of Fm' (dots) and of F (triangles) and thus ΔF. The values of ΔF/Fm' recorded in part B give an estimation of effective quantum yields ir, the presence and absence of CO_2, which is in fact the final electron acceptor.

Fig. 2.38. Study of electron flux in assimilating conditions (presence of CO_2) or non-assimilating conditions (absence of CO_2) (*Schreiber et al., 1995, in Schulze and Caldwell, 1995*).

exchanges under conditions of practically no CO_2 is analysed much better.

1) $2H_2O \xrightarrow{PII} 4e + 4h^+ + O_2$ Mehler reaction, causing qP

2) $4O_2 + 4e \xrightarrow{PSI} 4O_2^-$

3) $4O_2^- + 4H^+ \xrightarrow[SOD]{} 2H_2O_2 + 2O_2$

4) $2H_2O_2 + 4e + 4H^+ \xrightarrow[MDR]{APO} 4H_2O$ reaction of APO also causing qP

5) $2H_2O \xrightarrow{PSII} 4e^- + 4H^+ + O_2 \neq$

MDR = monodehydroascorbate reductase, and SOD = superoxide dismutase.

Monodehydroascorbate is the acceptor of electrons produced by reaction (5). It is used to characterize O_2.

The photoacoustic method can also be used (Reising and Schreiber, 1992). In fact, the H_2O_2 formed after the Mehler reaction must be considered a natural Hill reactive.

The reduction of monodehydroascorbate in light (reaction 5) is critical for the energization of the membrane and its implications for regulation and photoprotection.

In all circumstances in which the reduction of CO_2 is limited (i.e., before the activation of the Calvin cycle, or when the light is very intense, or if the stomata are closed, in case of drought, etc.), the reduction of O_2 and H_2O_2 serves as an exhaust valve for the electron flux, and it also liberates $H^+ \rightarrow \Delta pH$, which may lead to regulation of the dissipation of energy, and thus to the retroactive regulation of PS II activity, which is reflected in qN.

A cyclic electron flux around the PS I can contribute to ΔpH (Fig. 2.39). This is to be considered most of all in the systems that have much more PS I than PS II, for example, cyanobacteria and the sheaths of C4 plants. If PS I and PS II are overall in equilibrium, a cyclic transport of electrons by PS I is always possible. The electrons that will be in excess could go upstream and reduce the pool of plastoquinones (by the bias of plastocyanine and of the cyt Fb6 complex, probably). It thus seems that in any state in question there must be an equilibrium of the whole process by oxygen allowing excess electrons to escape. The application of the preceding methods in measurements of PS parameters in the field is obviously fundamental to ecophysiology.

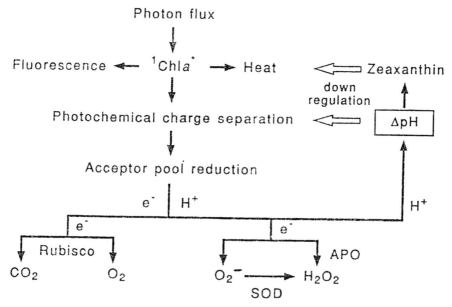

When the photosynthetic apparatus is exposed to a photon flux that exceeds the capacities of the Calvin cycle at a given moment, the great reduction of acceptor pools may lead to damage by photoinhibition. Therefore, alternative electron acceptors and a retroactive regulation of PS II are required. Photorespiration represents the major alternative sink to electrons. Moreover, the neutralization of the anion superoxide by the superoxide dismutase (SOD) and then reduction of H_2O_2 (Mehler reaction, ascorbate peroxidase, APO) cause the drop in ΔpH and the formation of zeaxanthin.

Fig. 2.39. The regulatory effects of oxygen concentration and transthylakoidal pH gradient on electron flux
(Schreiber et al., 1995, in Schulze and Caldwell, 1995).

4.1.6. *In situ measurement of $\Delta F/Fm'$ and rate of electron transfer*

Following the pulse saturation method, and the process of calculation of Genty et al. (1989), active efforts were made to develop portable fluorimeters for the field, to obtain $\Delta F/Fm'$ and $PFD \times (\Delta F/Fm')$. With the photoacoustic method we can measure Fo, Fo', Fm, Fm', and F, and from these we can deduce by computer Fv/Fm, $\Delta F/Fm'$, NPQ, Qp, and QN. We can also measure PFD (intensity of natural light).

Example: Photosynthetic activity in two bean leaves was analysed at noon (Fig. 2.40). One of the two leaves was treated for two hours after sunrise with 5 min. of heating at 43°C. The leaves were previously placed in the dark for 10 min. to obtain Fo and Fm. Then, continuously increasing illumination was applied directly on the portion of leaves in which fluorescence was analysed, from 50 to 1800 $\mu M/m^2/s$, with additional pulses each time allowing analysis of the variable fluorescence.

The heat treatment causes a drop in the photosynthesis, and even in the quantum yield at low light. Fo increases, Fm decreases, and Fm'

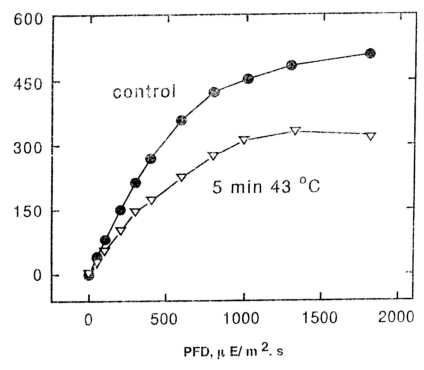

Fig. 2.40. Rapid measurement of a thermal effect by fluorimetry
(Schreiber et al., 1995, in Schulze and Caldwell, 1995).

decreases considerably even below 50 µM/m²/s, indicating a significant NPQ.

At 1000 µM/m²/s, the electron flux drops by 32%, which corresponds to the product of the variation of Fv'/ Fm' and of Qp, that is, to a reduction in the number of "open" centres and also in their efficiency.

4.1.7. Limitation of yield and intense light
The optimum yield indicated by Fv/Fm is 0.83, if it is considered that 84% of the incident light is absorbed and that there are two photosystems. From this is deduced a relative maximum rate of photosynthesis equal to the incident light × 0.83 × 0.84 × 0.5 = 0.35 × PFD. This value of 0.35 is close to that of the parameter m = $\varnothing_{P_0}$ (see section 4.1.4) derived experimentally from measurements by Weiss and Berry (1987).

If there is a lowering in photosynthesis under a given illumination (PFD), or $L_{(PFD)}$, then:

$L_{(PFD)}$ = 1 − ∆F/(Fm' × 0.83) or
$L_{(PFD)}$ = 1 − (Qp × [Fv'/Fm'])/0.83
L becomes equal to zero if Qp = 1 and Fv'/ Fm' = 0.83

L is in fact an integral measurement of the totality of the stress. It is more useful to measure if in intense light because it then integrates deficiencies at the level of the Calvin cycle or of storage in "sinks". Fluorescence thus has applications in the study of adaptation to intense light and of photoinhibition stress, and measurement of recovered activity (Bjorkman and Demming-Adams, 1995). It is also useful in analysis of the xanthophyll cycle (Fig. 2.41).

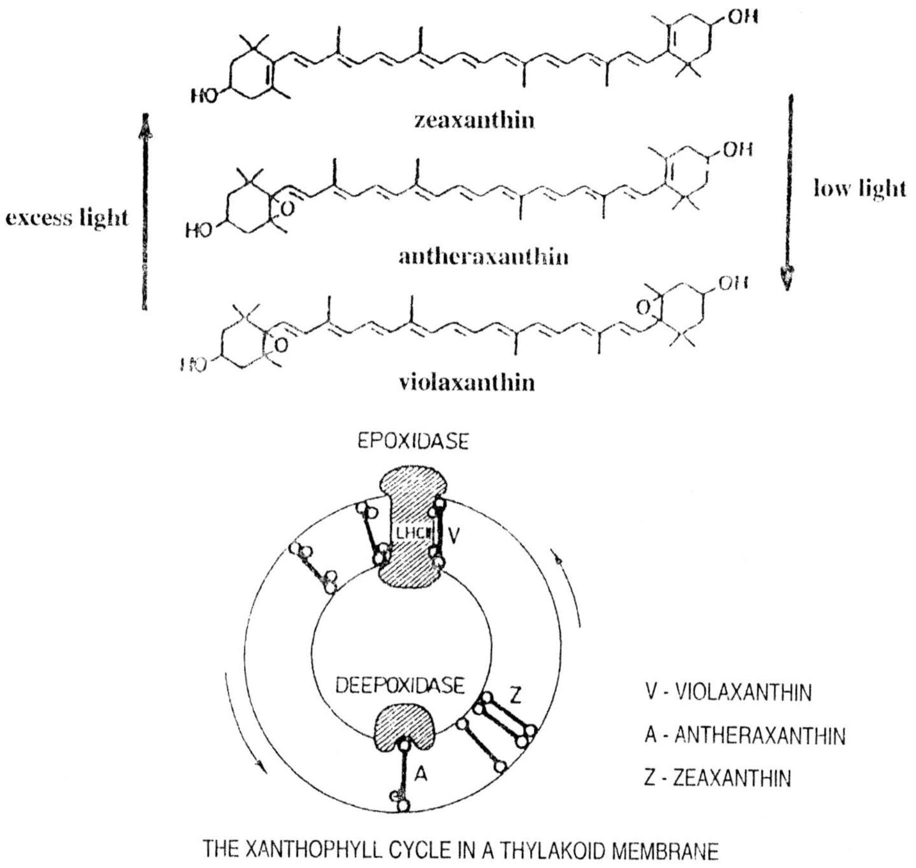

excess light

low light

zeaxanthin

antheraxanthin

violaxanthin

EPOXIDASE

LHC V

DEEPOXIDASE Z

A

V - VIOLAXANTHIN

A - ANTHERAXANTHIN

Z - ZEAXANTHIN

THE XANTHOPHYLL CYCLE IN A THYLAKOID MEMBRANE

Above are the three molecules involved with the direction of interconversions. Below is the thylakoid. The xanthophyll molecules occupy the hydrophobic core of the membrane, which is about 3 nm (which corresponds to the length of the molecules). The epoxidation is produced in contact with LHC (antenna complexes) that in the presence of NADPH have an epoxidase activity of enzymatic nature. Deepoxidase is close to the lumen. Lateral mobility of molecules is needed, as well as a "flip-flop" motion, since the two epoxides are at each end of the molecule.

Fig. 2.41. The xanthophyll cycle
(*review of Gruszecki, 1995*).

Environmental factors such as temperature and dryness must be well understood and controlled in order to draw valid conclusions on the "photosynthetic health" of a plant from fluorescence data.

4.2. Application to the study of photoinhibition

Along with the kinetics of fluorescence, Bjorkman and Demming-Adams (1995) studied the regulation of light interception, regulation of conversion of energy, and non-radiative dissipation. They also gave some examples of protection against intense light, for example in *Encelia farinosa*, a shrub of the American semi-desert. In the relatively cold season these shrubs have few hairs, but in the hot season there are many hairs that are larger and denser and charged with air. In Death Valley, this plant absorbs only 29% of the visible light in summer. Another species, *Atriplex hymenelytra*, which also grows in Death Valley, is subjected to intense light reflected from the soil, because of the presence of salt deposits. Morphological mechanisms of the epidermis help the plant avoid photoinhibition. In summer, 60% of the light is reflected, but in winter only 30% of the light is reflected.

Fluorescence can only dissipate 3 to 4% of the received light energy, but it can provide information about basic photosynthetic yield and photorespiration. A major part of the energy undergoes non-radiative dissipation, i.e., in the far infrared.

4.3. Conclusions

The kinetics of fluorescence induction was at first used mostly to improve the basic analysis of photosynthesis. Now it appears valuable in measuring productivity and stress situations in the field with equipment designed for the purpose. In a few seconds, the potential maximum quantum yield of PS II can be recorded (Fv/Fm), as well as the effective yield in a given situation ($\Delta F/Fm\phi$), the effective rate of electron transfer, the efficiency of photon capture (Fv'/Fm'), photochemical and non-photochemical quenchings (Qp and Qn), as measurements of the number of open PS II centres, and regulation of the use of captive energy. One may also add $L_{(PFD)}$, the lowering of photosynthesis in given environmental conditions. The methods are non-destructive and the same part of the leaf can be followed during the course of its development and during a daily cycle.

In future we can quickly test new plant varieties, possibly obtained by genetic engineering, for their response to the environment. In a stress situation we often see non-photochemical quenching first of all, which can immediately be detected by fluorescence. It is a response of the plant to avoid photoinhibition, which involves the capture of electrons by oxygen, the xanthophyll cycle, the pH gradient, and the controlled inactivation of centres.

5. RESPIRATION OF HIGHER PLANTS AND ITS RELATIONSHIP TO PHOTOSYNTHESIS

5.1. Introduction

Nocturnal respiration and its variations according to the ecology of the species are well known (Table 2.9), but there is also respiration during the day. What is the quantity of CO_2 that is released by respiration, in parallel with photosynthetic activity? Few detailed studies have been done because physiologists have rarely measured the rate of respiration on entire plants (Montieth, 1972). It is also useful to measure the energy yield of respiration. In addition, we must consider respiration in situations of growth, maintenance, and stress (Amthor, 1993).

5.2. Routes and control of respiration

Glycolysis, the pentose "cycle", the Krebs cycle, and phosphorylation are all involved in the process of respiration. Glycolysis is the most universal of the metabolic routes. Part of the pentose cycle and the Krebs cycle probably appeared later in the evolution of plants. The appearance of photosynthesis with two photosystems has permitted oxidation of iron salts and then emission of atmospheric oxygen, and probably the completion of the Krebs cycle with α-aceto-glutarate dehydrogenase activity (Gest, 1987).

Mitochondria and chloroplasts appeared in the wake of endosymbiotic phenomena (Sitte and Eschbach, 1992).

A certain number of properties are specific to plant mitochondria, and a complete description has been given by Amthor (1995) (Fig. 2.42). It may

Table 2.9. Respiration of adult leaves (representative of basal metabolism) in summer at 20°C, early in the night (results from many authors, in Larcher, 1995)

	$\mu mol/m^2/s$
Cultivated plants	2-6
Wild plants in the sunlight	3-5
Wild shade plants	1-3
Holoparasites	3-6
Tropical trees	
Leaves in sun	0.3-0.5
Leaves in shade	0.05-0.2
Caducifoliate trees Leaves in sun	1-2
Leaves in shade	0.2-0.5
Trees with persistent leaves (temperate zone)	
Leaves in sun	0.8-1.4
Leaves in shade	0.2-0.5
Alpine Ericaceae	0.3-1

Note: Mitochondrial respiration does not differ significantly between C3 and C4 plants (Byrd et al., 1992). It is generally observed that respiration is higher early in the night than near the end of the night (see, for example, Leclerc and Blaise, 1991).

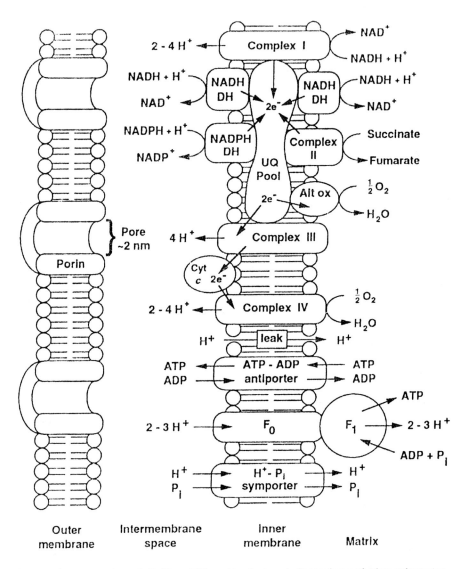

The membrane complexes I, II, III, and IV are identical or similar to those of other eukaryotes. An NADH dehydrogenase located on the matrix side is different from complex I and insensitive to rctenone. It does not exist in animals. An alternative oxidase (alt OX), very active in certain plants, donates a part of the electrons of the UQH2 pool directly to oxygen. NADHH[+] and NADPHH[+] (in plants and mushrooms) as well as ATP pass freely through the pores of the outer membrane.

Fig. 2.42. Respiratory chain of mitochondria of higher plants
(Amthor, 1995, in Schulze and Caldwell, 1995).

be noted, for example, that succinate-CoA ligase phosphorylates ADP rather than GDP. There is also a transport of electrons resistant to rotenone, and most of all a cyanoresistant respiration. Transport of oxaloacetate (OAA), oxidation of malate in the presence of OAA and absence of pyruvate, as well as the presence of certain cytochromes are specific to plants. In relation to the phenomena of photorespiration, glycine decarboxylase is abundant. Finally, the genome is larger and more complex than in other living things (Douce and Neuburger, 1989).

5.2.1. Control of respiratory rate
In rapidly growing cells, the quantities of substrate or enzymatic complexes can be limiting. In adult or slowly growing cells, there is certainly short-term control (a few seconds to a few hours) by means of the abundance of products, mostly ATP. Compared with photosynthesis, the energy efficiency of heterotrophic metabolism, particularly respiration, is impressive.

5.2.2. Energy conservation during respiration
About 30 to 36 ADP can be phosphorylated when a hexose is completely oxidized. The rotenone-resistant route is used when NADH is in excess (short circuit by complex I). The alternative cyanoresistant route is very common in plants, and it is used when UQ is greatly reduced. NADH is oxidized even if ADP is low, if the proto-motor force is high, and a passive dissipation of proton gradient is produced. The same process occurs for UQH_2.

5.2.3. Respiratory rate and level of glucides
The relationship between respiratory rate and level of glucides may be very clear in growing tissues but absent in adult organs. The respiration during the growth is thus controlled by the level of carbohydrates, but the respiration of maintenance is not (Amthor, 1989). For example, an input of sugars may stimulate the absorption and assimilation of nitrates by the roots, and this nitrogen input will perhaps stimulate cell division and differentiation, which consume high levels of energy.

An increase in sugars may lead to an increase in the (potential) respiratory capacity (Farrar and Williams, 1991). Sugars can simultaneously act as messengers as well as substrates. One consequence of their messenger nature may be not only the stimulation of respiratory power but also the inhibition of the expression of certain genes of the photosynthetic mechanism (Schafer et al., 1992).

5.2.4. Respiration of leaves during photosynthesis
It must always be remembered that photosynthesis, respiration, and photorespiration occur simultaneously. However, their sensitivity to temperature is different, and this sensitivity is involved in the formation of

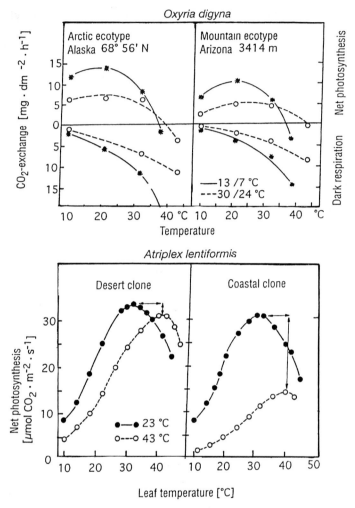

Two examples are presented. *Oxyria digyna* is distributed in the northern hemisphere in the Arctic zone and high mountains of the temperate zone (it is an Arcto-alpine plant). Two ecotypes were studied in two pairs of day/night temperature. Respiration quickly became very high when measured at high temperature (close to 30°C) in plants that were habituated to a cool temperature (13°/7°C) close to the normal situation. Photosynthesis of plants at 13°/7°C was better than those of plants at 30°/25°C, but not beyond 30°C. The Arizona ecotype could to some extent adapt itself to a hot climate. *Atriplex lentiformis*, in the southwestern United States, presents a coastal ecotype bordering a cold ocean and a hot desert ecotype. After cultivation in ambient temperature going up to 43°C at midday, the maximum of photosynthesis moved from around 10°C towards the heat (horizontal arrows) until it reached around 40°C. Generally, however, photosynthesis of the coastal ecotype decreased greatly when it was cultivated in hot conditions, unlike what was observed in the other ecotype.

Fig. 2.43. Acclimatization of photosynthesis and respiration to changes in temperature *(data are from Billings et al., 1971 for Oxyria and Pearcy in Osmond et al., 1980 for Atriplex in Larcher, 1995).*

ecotypes having more or less ecological amplitude (Fig. 2.43). The sensitivity is also different in the face of hydric deficits leading to low water potentials (Fig. 2.44).

In leaves under normal conditions, without excessive light, the CO_2 assimilation is 10 to 20 times as rapid as the CO_2 emission in dark. The chloroplast in light directly provides reductive power to cytosol and indirectly provides ATP by the export of dihydroxyacetone phosphate. A relation of competences between the mitochondria and the chloroplasts is worth studying (Fig. 2.45). It must be noted that the needs of mitochondrial respiration are less important in light, the cytosolic ATP/ADP ratios being high.

Moreover, it may be considered that, in light, the glycolysis and the pentose route no longer function in the chloroplast. According to Kromer and Heldt (1991), in very strong light NADP may be much superior to

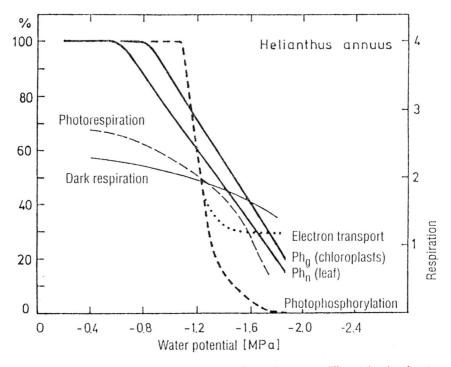

The studies were conducted on dehydrated sunflower leaves at different levels of water potential. The activities are in percentage of activity of well-irrigated leaves. Photosynthesis was rapidly affected. Photophosphorylation was affected at lower potential but the impact was harsh. The relative resistance of mitochondrial respiration to low water potential was noted.

Fig. 2.44. Different sensitivity of photosynthesis and respiration during hydric deficit *(Boyer and Bowen, 1970; Keck and Boyer, 1974; Ortiz Lopez et al., 1991 in Larcher, 1995).*

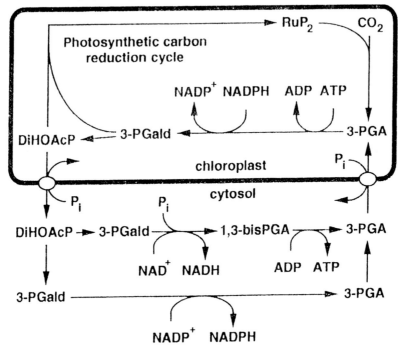

Dihydroxyacetone phosphate (DiHOAcP) is exported (antiport with Pi). The conversion of 3 phosphoglyceraldehyde (3PGald) into 3 phosphoglyceric acid (3-PGA) enables the formation of ATP, by the intermediary of 1,3-diphosphoglyceric acid, and the 3-PGA returns to the chloroplast. If the cytosolic Pi is low, however, the phosphoglyceraldehyde is directly transformed into PGA, with formation of NADPHH+.

Fig. 2.45. Transports associated with the influence of chloroplast on the increase of energy charge of cytosol
(Amthor, 1995, in Schulze and Caldwell, 1995).

NADH in the foliar cytosol, which will not block the tricarboxylic cycle (NADPH being used by the respiratory chain, as well as NADH), but will block the pentose routes.

Because of this simultaneity, the "exact" values of photosynthesis, respiration, and photorespiration cannot be assessed by simple techniques of CO_2 and O_2 measurement or even, for various reasons, by isotope analysis. In the case of photorespiration, however, a solution may be found with the ratio of carboxylase activity and oxygenase activity of Rubisco (Kirschbaum and Farquhar, 1987).

The Kok effect, which is manifested by an abrupt fall in the apparent quantum yield of photosynthesis, close to the light compensation point (Kok, 1948), expresses a partial inhibition of respiration by light (Sharp et al., 1984). Sometimes, the Kok effect is not observed, which could result either from a too low concentration of O_2 or a high concentration of CO_2 (Amthor et al., 1992) having from the beginning inhibited respiration. The degree of inhibition can be calculated in normal conditions (Leclerc, 1981) or in conditions close to normal according to Brooks and Farquhar (1985).

From the three stages of respiratory catabolism it appears that cytosolic glycolysis would be little affected by light. Inhibition of the tricarboxylic cycle would also be partial. The need to provide carbonate skeletons could explain its maintenance. As regards oxidative phosphorylation, it could be noted that in leaves that do not easily produce amidon, respiratory ATP would be necessary for rapid synthesis of cytosolic saccharose-phosphate (Hansson, 1992). Moreover, mitochondrial activity, by massively using the reductors produced by photosynthesis, could reduce (Saradadevi and Raghavendra, 1992) the phenomena of photoinhibition.

Photosynthesis-respiration ratios that are apparently still unclear are explained overall by the diversity of the physiological stages of plants, or by the part of foliar respiration in the general respiration of a plant or a tree.

5.2.5. Photorespiration and mitochondrial metabolism

It must be recalled here that many types of oxidation, some of which are realized in the peroxisomes in the process of photorespiration, are extramitochondrial (Table 2.10) and occur with relatively low affinities.

The glycine decarboxylase activity that produces CO_2 and NADH plays a central role. If the plant is placed in conditions of high presence of glycine, and thus maximal activity, the addition of malate increases the consumption of O_2 (Wiskich et al., 1990). In this manner the Krebs cycle, which is the major provider of oxidative photophosphorylation, continues to function.

It must be ensured that in the mitochondria the enzymes are more or less located on the membranes and that thus the Krebs cycle, which

Table 2.10. Extramitochondrial oxidation, produced by some oxidases
(data from several authors)

Enzyme	EC	Km (O_2), mol/l, at 25°C	Prosthetic group
Ascorbate oxidase	1.10.3.3	3×10^{-4}	Cu
Phenole oxidase	1.10.3.1/2	close to 10^{-5}	Cu
Urate oxidase	1.7.3.3	close to 10^{-4}	Cu
Glycolate oxidase	1.1.3.1	close to 10^{-4}	FMN
O-aminoacid oxidase	1.4.3.3	2×10^{-4}	FAD
Fungal glucose oxidase	1.1.3.4	2×10^{-4}	FAD
Fungal ethanol oxidase	1.1.3.13	?	FAD
CCO	1.9.3.1	close to 10^{-7}	hemes with Cu, Fe

The concentration of dissolved oxygen in water at 25°C is 2.6×10^{-4} mol/l, i.e., close to Km of many extrachloroplastic oxidases. The values of Km are much greater than those of the cytochrome C oxidase complex (CCO), indicated for comparison, and thus the affinities for oxygen are weak.

demands a great deal of NAD^+ and produces a great deal of $NADHH^+$, may be quite independent of glycine decarboxylate activities and of transport of OAA and malate linked to the functioning of peroxisomes and using also a pool of NAD^+ and $NADHH^+$.

5.2.6. Links between photosynthesis during the day and respiration during the night

Photosynthesis plays a positive role in respiration during the night (Fig. 2.46, Leclerc and Blaise, 1994). This is very likely due partly to the energy costs of various transports such as charge of phloem, discharge of phloem, and charge of vacuoles in the reserve organs (Amthor, 1993). Getz (1991), for example, clearly demonstrated saccharose accumulation, dependent on ATP, in the vesicles of tonoplast of beetroot.

If a high rate of photosynthesis prevails, one consequence could be the stimulation of alternative respiration, which allows the rapid regeneration of NAD^+. It is also known that equilibrium between the different metabolites changes according to whether the change from day to night is abrupt or progressive. Finally, not much is known about the relations existing in large trees, where the great distance between sources and sinks must lead to considerable phenomena of cushioning.

The direct effects of blue light on respiration, independent of photosynthesis, are known. The effects persist in the darkness that follows (Kowallik, 1982). According to Sims and Pearcy (1991), respiration is

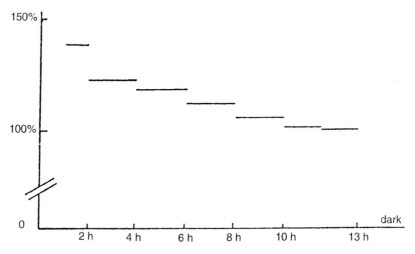

X-axis, night temperature. Y-axis, respiration intensity, 100% representing the value at the end of the night. From the first to the second hour of the night, the coefficient of variation of measurements was 20%, and this coefficient fell to 5% by the end of the night.

Fig. 2.46. Variation of nocturnal respiration
(Leclerc and Blaise, 1994).

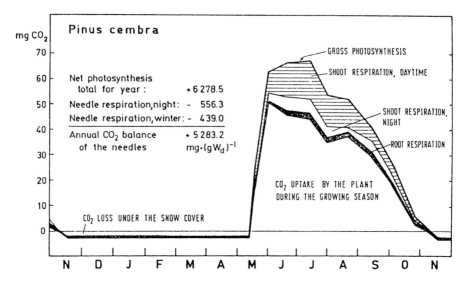

Young plants of *Pinus cembra* living in the upper subalpine zone were studied. The carbon loss (by respiration) was not entirely negligible during the winter. During the summer, the respiration of aerial parts during the day (hatched part) is high because of the photorespiration. The other components of loss in summer are root respiration (black part) and the nocturnal mitochondrial respiration of plant needles. Other results on several tree species are found in Tselniker (1993).

Fig. 2.47. The annual carbon budget in trees
(*Tranquillini, 1959, in Larcher, 1995*).

linked to real photosynthetic activity of the preceding days rather than the photosynthetic capacity.

The effects of CO_2 concentration have been studied extensively, partly, of course, because of present increases in atmospheric levels of CO_2.

High CO_2 levels often lead to an increase in the amidon/saccharose ratio, saccharose being more directly linked to respiration than amidon. This is perhaps one explanation for the inhibitor effect of CO_2 on respiratory activity. It must be noted that although the often long-term increase in CO_2 leads to a decrease in respiration estimated on the basis of dry biomass, the effect is less evident on the basis of nitrogen or proteins (Amthor, 1991).

A high ambient CO_2 level leads to a decrease in photorespiration, of course, and the lower supply of NADH to the mitochondrion could in the short or long term lead to a decrease in the oxidative machinery of the mitochondrion. There is also a direct effect of high nocturnal CO_2 levels on respiratory activity. The inhibitor effect is reversed if the CO_2 level falls (Amthor et al., 1992).

High nocturnal CO_2 levels diminish the mobilization of foliar amidon and, the simple oses being rare, also diminishes the respiration (Wullschleger et al., 1992).

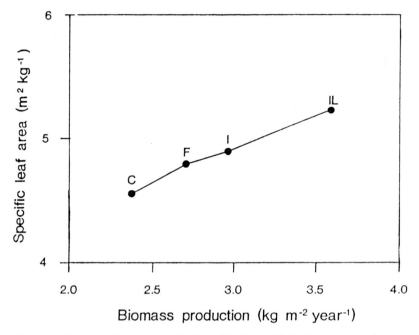

In a *Eucalyptus* the leaf index was varied with fertilization and irrigation, and a correlation was clearly seen. The slope also shows that the various cultivation treatments had clearly more effect on production than on the leaf index. C, control. F, fertilized. I, irrigated. IL, fertilized and irrigated.

Fig. 2.48. Production of biomass and leaf index
(Pereira, 1995, in Schulze and Caldwell, 1995).

5.2.7. Photosynthesis and root respiration

Root respiration is often studied in solution and rarely *in situ*. One consequence of solutions is that certain elements may "leak away" in abundance, unlike what happens in soil, where the external liquid phase is greatly reduced. The equilibrium in a nutritive solution may thus be very different from what it is in the soil.

On herbaceous plants we can see, in a few hours following a change in light, a change in the root respiration (Massimino et al., 1981). Recent photosynthates would have a much clearer influence on the root respiration than reserve photosynthates (Ryle et al., 1985).

These recent photosynthates, via respiration, facilitate the assimilation of NO_3^-, K^+, etc., but this is seen mostly in laboratory experiments.

5.3. Conclusions

The interactions between photosynthesis, respiration, and photorespiration are evident and result in fixed quantitative ratios between overall

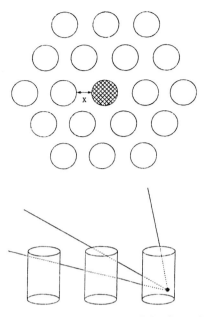

Each plant can be compared to a cylinder capturing light. Several distances x can be used between individuals, and individuals may also partly overlap. Photosynthesis of a typical plant is calculated taking into account its own absorption and the light absorbed by its neighbours at different directions in space.

Fig. 2.49. Extrapolation of results from one plant to a population
during a modelling experiment
(Beyschlag et al., 1990, in Schulze and Caldwell, 1995).

incorporated carbon, carbon used for growth, and carbon lost during respiration for a given plant species (Farrar and Williams, 1991). However, the results and regulations of these interactions are not sufficiently understood, particularly in the medium term.

In a complex environment, the consequences of stress on the whole plant may be more important than might be judged from the impact on photosynthesis alone.

6. OVERALL PRODUCTIVITY ON THE ECOSYSTEM SCALE

In ecosystems, the plant reacts to its neighbours and its consumers. The result is manufacture of more tissues for maintenance, hardening, and synthesis of defensive and other substances, which affect the general productivity (see Table 1.11).

The plant reacts individually and as part of a population to the abiotic environment. The result is the realization of an annual energy budget

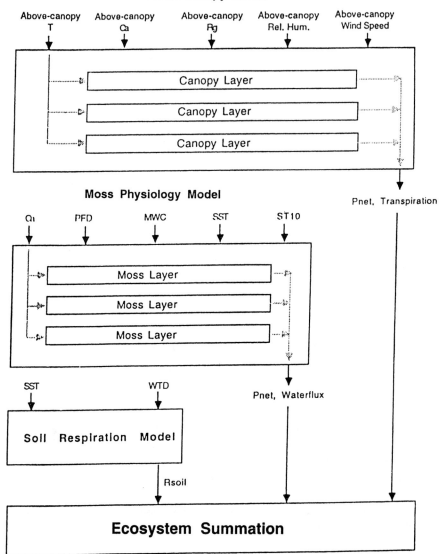

The system presented can be applied to the tundra or the boreal forest. The different layers of the canopy and layers of mosses and sphagnum are distinguished. Physical data taken above the canopy are introduced: temperature (T), concentration of CO_2 (Ca), light (Rg), relative humidity of air, and wind speed. At the level of mosses the CO_2 concentration (Ca) is introduced, the light that is passed across the canopy (PFD), the moss water content (MWC), the surface soil temperature (SST), and soil temperature at a depth of 10 cm (ST10). The rates of net photosynthesis and of transpiration are predicted. At the level of soil the CO_2 emission is estimated from SST and soil water content (WTD).

Fig. 2.50. Model of carbon balance of a plant community
(Tenhunen et al., 1995, in Schulze and Caldwell, 1995).

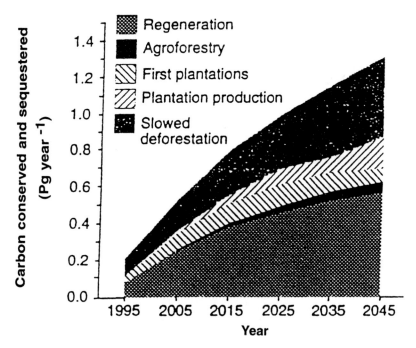

In this simulation, a model valid for low latitudes, the slowing down of deforestation will have a highly significant impact on the trapping of CO_2. A significant impact can also be achieved if existing forests that were more or less degraded are managed in order to enhance their natural regeneration.

Fig. 2.51. Quantity of carbon that could be trapped in the future as a function of forest policies
(Dixon et al., 1994).

(Fig. 2.47) and also a budget for a longer term. Depending on the complexity of the architecture in an ecosystem, a forest for example, there are also relationships between productivity and leaf index (Fig. 2.48). We can also see in the framework of forest plantations the incidence of irrigation and fertilization.

After measurements of photosynthesis at the level of the leaf or the whole plant, an attempt is made, of course, to construct models, to extrapolate from an individual (Fig. 2.49) to a population. The models extend to plant communities (Fig. 2.50), considering the different taxonomic and biological groups. An essential objective of these models is, with estimates of present productivity, to forecast future productivity, for the particular purpose of trapping CO_2 by vegetation (Fig. 2.51), which is the key to the control of climatic change.

Chapter 3

Hydro-mineral Equilibrium

1. PLANTS AND WATER

1.1. Introduction

The water content of plants varies greatly according to the species, the organs, and the environment in which they live. A lettuce may contain 90 to 93% water, and a leaf often has 80 to 90% water, but freshly cut wood may have 30 to 50% water. In certain situations, high water content may be accompanied by high salt content or content of soluble organic matter inducing a noticeable drop in the water potential, which could come down to –5 or even –8 MPa in halophytes of the temperate regions.

Certain plants may tolerate considerable reduction in water content, as with mushrooms, sub-aerial algae, mosses, and lichens. The cells of these organisms have only small vacuoles, and the fine structure of their hyaloplasm is not destroyed by the loss of water. In case of desiccation their water potential decreases drastically, even if there are only a few mineral ions, but there will then be dissolved organic substances, concentrated, or structures that can absorb water.

Stenohydric plants have only a narrow variation of water content, but they could be euryhaline, that is, they could have a salt content that is possibly very high, or stenohaline, or oligohaline, tolerating only low levels of dissolved salts, and poikilohydric (or euryhydric) plants with a widely varying water content, particularly dropping a great deal, whether they are euryhaline or oligohaline.

The water content of terrestrial plants is the result of an equilibrium between what is lost by evapotranspiration, often essential to the entry of CO_2, and what enters by root absorption, which depends on the water potential of soils as well as the stem conductance capacity.

Finally, the need for water in plants is itself variable. Of course, all plants require water, since it is a substrate as well as support of metabolic reactions. But a certain number of key enzymes may, depending on the species, tolerate low water potential.

Thus, we must examine what enters, what circulates, and what leaves the plant, and the hydric equilibrium that can be compatible with active life or inactive survival, all in relation with the hydric characters of the environment. In the discussion on equilibrium we will of course examine cellular ecophysiology (Fig. 3.1) in higher plants as well as in thallophytes and prokaryotes.

Moreover, we must not limit ourselves to the study of the individual. We must, in the context of complex interactions among plants, in a corn field or in

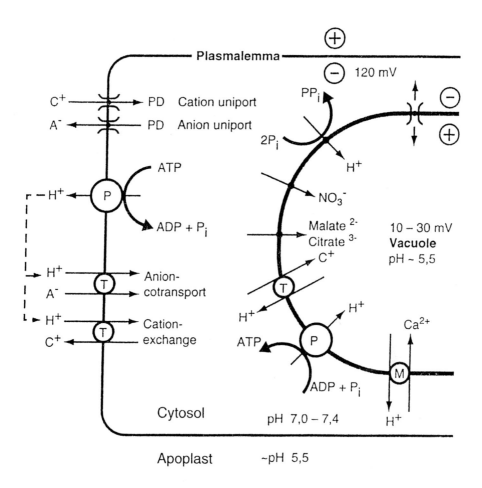

Fig. 3.1. Diagram of transport across the plasmalemma and the tonoplast *(Pitman and Luttge, 1983; Kaiser et al., 1988; Martinoia, 1992).*

a forest, analyse the problem of water on the scale of plant communities (Tables 3.1 and 3.2), in relation with the environment.

Table 3.1. Annual and daily transpiration of plant formations
(data from several authors, in Larcher, 1995)

Type of vegetation	Transpiration of the biocenose	
	mm/year	mm/day
Ligneous formations		
Plantations of tropical trees	2000 to 3000	
Tropical rain forests	1500 to 2000	
Deciduous temperate forests	500 to 800	4 to 5
Conifer forests	300 to 600	2.5 to 4.8
Sclerophyte forests	400 to 500	
Wooded steppes	200 to 400	
Heathland	100 to 200	2 to 5
Prairies and other herbaceous formations		
Sedges and juncus	1300 to 1600	6 to 12
Megaphorbs	800 to 1500	
Humid prairies	1100	
Fields planted with cereal crops	400 to 500	
Prairies and savannahs	4 to 6	
Mesophile pastures	300 to 400	
Low steppes	around 200	0.5 to 2.5
Vegetation in open environments		
Halophyte communities	2 to 5	
Mountain ridges	10 to 20	0.3 to 0.4
Lichens of the tundra	80 to 100	
Barren deserts	0.01 to 0.4	

1.2. Water that enters

1.2.1. By direct absorption

The case of thallophytes can be compared to that of cormophytes. In thallophytes, liquid water is directly absorbed by the aquatic species. In a terrestrial environment, it is also absorbed by capillary action and saturation from rain, from wet soil surfaces, or from dew or fog. A fruticolous or foliate lichen may multiply 3 or 4 times, and a gelatinous lichen (e.g., *Collema*) may multiply 10 to 15 times. The absorption of liquid water is rapid, a few minutes to about an hour. Certain species, particularly the Chlorophyceae (e.g., *Trentepohlia*), can also absorb water vapour, provided the relative humidity of the air is very high. But the absorption of water vapour is slow. Mosses behave like thallophytes. Certain vascular plants (in principle made "impermeable" by their cuticle or cork layer) are capable of absorbing water vapour by means of special structures: e.g., permeable openings in the epidermis (hydathodes), base of certain trichomes, scales of epiphyte Bromeliaceae, or particular stomata of aquatic plants.

Table 3.2. Hydric equilibrium of plant formations (Duvigneaud, 1967; Stanhill, 1970; Mitscherlich, 1970; Grin, 1972; Doley, 1981; Brunig, 1987)

Type of vegetation	Region	Annual precipitation	Evapotranspiration, % of precipitation	Drainage (surface and ground water), % of precipitation
Forests				
Primary tropical rain forest	N. Australia	3900	38	62
Tropical rain forest	S.E. Asia	2000–3600	50–70	30–50
Deciduous tropical	S.E. Asia	2500	70	30
Deciduous temperate	Central Europe	600	67	33
Coniferous	Central Europe	730	60	40
Mountain	S. Andes	2000	25	75
	Alps	1640	52	48
	Central Europe	1000	43	57
	N. America	1300	38	62
Prairies				
Savannahs	Tropical	700–1800	77–85	15–23
Marsh	Central Europe	800	>150	
Pasture land	Central Europe	700	62	38
Alpine pastures (in vegetative season)	Europe	500–600	25–40	60–75
Steppes	E. Europe	500	95	5
Open formations				
Arid mountainous prairies	N. Argentina	370	70–80	20–30
Semi-deserts	Subtropical	200	95	5
Absolute deserts	Subtropical	50	>100	0
Tundra	N. America	180	55	45

1.2.2. Through roots

1.2.2.1. Soil granulometry

Rainwater that falls on the soil percolates (circulates by gravity) more or less rapidly according to the soil granulometry. If the elements are coarse (sand), water percolates some centimetres or millimetres a day, but in clayey soils it percolates only a few millimetres a day. Part of the water is retained if the soil is rich in alluvium. A fairly large quantity of water is retained especially if the particles are smaller than 10 to 20 µm. The **field capacity** of soils is defined as the percentage of water that can be retained, once the percolation ends, in g of water that can be retained per 100 g of dry soil. This capacity is very high in soils with fine elements and/or with a large proportion of colloids. If the rainfall, or snowfall, is prolonged, a good part of the water that percolates may be absorbed through the roots.

Much of the water is not free, it is retained by capillary force. Capillary potential of soil water can be calculated from the following formula:

$\Psi_{cap} = -4\sigma/d = -290/d$, in J/kg, with d in µm; and $\Psi_{cap} = -0.29/d$ MPa $= -2.9/d$ bar, σ being the water surface tension, and d the diameter of pores in microns.

In a soil rich in alluvium, the range of granulometry is vast, and in the beginning the water is practically not retained. Once the reserves retained between the large grains disappear, some water remains between the small grains and that is more difficult to extract. If d becomes less than 0.1 to 0.2 μm, depending on the species, the mesophytes can no longer absorb it.

1.2.2.2. Conditions of water absorption by roots

The rate of absorption of water by a plant, V_{abs}, depends on the absorptive area of the root system, A, the difference of water potential of water contained in the soil, Ψ_{soil}, and the water contained in the root, Ψ_{root}. The water enters the plant if $\Psi_{root} < \Psi_{soil}$, and the rate is inversely proportionate to the resistance to transfer in the plant (for example, formation of cork) represented by ΣR.

$$V_{abs} = A(\Psi_{soil} - \Psi_{root})/\Sigma R$$

The root establishes a deficit of water potential, of a few tenths of an MPa, which is sufficient to draw much of the capillary water. In very dry soils, xerophytes may have root water potential as low as –6 MPa. The water that can be easily extracted, which comes from between fairly large grains, is quickly exhausted near the roots, and it may be replaced by lateral migration of soil water coming from zones further away, but it is a fairly slow transfer and, if too much water is exhausted, for example, during a hot day, the soil, which is very sensitive, shows resistance to transfer of water. In sandy soil, in particular, the water supply to the plant could be disrupted. The temperature gradients of soil also intervene. During the night, the deep (and moist) layers are warmer and the water vapour that they release condenses in the superficial cold layers, where it can be recovered by roots. In the middle and end of the day, another factor may bring water to the upper soil layers, which are hot: during the night, plants with deep roots pump water, which rises in the stem but also in the superficial roots, and during the day, these superficial roots release water into the surrounding soil if the water potential of the soil falls too much. The water thus liberated is not all that "free", but a neighbouring species of plant having only a superficial root system and a sufficiently low water potential may benefit from it, so this is important for competition between plants.

From one species to another, the root architecture will depend on four major variables: the capacity of the plant to tolerate low potentials, the soil texture, the location of capillary water or water circulating at depth, and the frequency and intensity of precipitation. Thus, many species of cacti have a superficial and extensive root system, in order to recover as much water as possible during rare rainstorms.

Other species of arid regions may combine a system of well-developed lateral roots on the surface with another deeper root system if there is an alluvial zone enclosing water reserves, and possibly with another still deeper system in contact with the water table. The growth of a root system follows a seasonal evolution of water levels in the different soil layers. The roots may die off in one place and at the same time develop and branch out in another place. A plant may draw water in increasingly finer capillaries, until in the deepest water-containing layers the potential of this soil water descends to the point of permanent withering, that is, at the lowest water potential that the plant can tolerate without dying. If the conditions are not extreme, the water potential observed in the leaves at the end of the night roughly corresponds to that in the overall soil area in which the roots are found. This last consideration raises the question, how much water circulates from the roots to the leaves, as a function of the gradient of the potential?

1.3. Water that circulates

1.3.1. Introduction

The movement of water in the plant is analogous to that of electrons in an electrical circuit. The differences in water potential are analogous to differences in the electric potential. The resistance to water circulation is analogous to electrical resistance. Generally, resistance in the plant is low, but resistance outside the plant (at the level of the cuticle + stomata system) may be very high. Also, we must consider the potential of gravity Ψ_g, which is due to the height of the water column in the plant.

At a given point on the plant (Richter, 1972), we have $\Psi = \Psi_{soil} + \Psi_g + \Sigma R \cdot Fl$. Fl represents the water flow between the soil and the point studied, and R represents the resistance of transfer between the point studied and the soil. R must be very low for a large flow of water to circulate, while keeping a slightly negative Ψ at all points of the plant, which implies that the stomata must be wide open. In general, Ψ decreases from the base to the top of the plant, except in particular cases such as plants that essentially absorb vapour from fog in certain climates. Over large distances, water moves en masse, by the flow of sap in the xylem vessels. Depending on the plant, the vascular system is more or less developed, which brings us to study particularly the rate and flow of sap.

1.3.2. The flow of sap

The flow (D), which can be expressed, for example, for a tree in litres per minute, depends on the difference between the foliar potential and the root potential, and the resistance R (or its inverse, conductance C) of the vascular system. V is the velocity of the current.

$D = V/R = C \cdot V$. C depends on the area of the section conducting water, A.

$D = Cst \cdot V \cdot A$, in kg/min. Cst depends on the length L of the conduction circuit and the specific hydraulic conductivity that is proper to each species and in a given environment, and it depends essentially on mechanical properties of vessels.

Specific hydraulic conductivity (Table 3.3) may be expressed in $cm^2/s/MPa$ = Cs. The rate of ascension can also be measured (Table 3.4), and it can be expressed in units of pressure.

$$D = V \ (MPa) \cdot A \ (cm^2)/L \ (cm)\text{-}Cs \ (cm^2/s/MPa)$$

If, for example, for a young angiosperm tree, we have V = 0.5 MPa, A = 50 cm^2, L = 500 cm, Cs = 100 $cm^2/s/MPa$, it can be seen that:

$$D = 0.5 \times 50/500 \times 100 \ cm^3/s = 5 \ cm^3/s$$

The area of conductance A is present in angiosperm trees in the form of an external ring of wood (sapwood), since the central part of the wood (the core) does not conduct. It is mostly the youngest part of the sapwood (1 to 3 years) that conducts. Moreover, the conductivity changes at different levels of the trunk and branches, tending to diminish towards the extremities.

The rate of flux is dependent on the flow D, and the section of conduction A, V = D/A, in the preceding example, V=1 mm/s=3.6 m/h. V is high (and thus also D) in the young parts of the sapwood, for example in oak. It may be interesting to compare A and the leaf surface Sf that will evaporate the water arriving by A. It is the relative area of conductance, or Huber value (Huber, 1956). In place of Sf, the weight of

Table 3.3. Hydraulic conductivity of xylem (Berger, 1931; Huber, 1956; Zimmerman and Brown, 1974; Raven, 1977; Ogino et al., 1986; Losch, 1990, in Larcher, 1995)

Plant material	Conductivity ($cm^2/s/MPa$)
Conifer wood	5 to 10
Ericaceae wood	2 to 10
Wood of evergreen trees with microporous xylem	3 to 15
Wood of deciduous trees with microporous xylem	18 to 50
Species with macroporous wood	100 to 350
Wood of creepers	300 to 500
Wood of roots in deciduous trees	200 to 1500
Fibrous roots	1 to 2
Vascular bundles in herbaceous plants	30 to 60

Table 3.4. Rate of sap ascension (data from several authors, collected by Larcher, 1995)

Plant groups	Rate (m/h)
Conifers	1 to 2
Sclerophyte trees and shrubs	1.5 to 3
Trees of tropical rain forests	18 to 34
Trees with microporous wood (temperate zones)	1 to 4
Trees with macroporous wood (temperate zones)	20 to 45
Herbaceous plants	10 to 60
Creepers	150

leaves to be supplied, FW_1, can be measured. We thus have A/FW_1. This value is very low in sciaphytes (around 0.2 mm^2/g), which lose little water and are habitually in a very humid atmosphere, but it is high in desert plants (around 2 to 3 mm^2/g), which lose little water but live in a very dry atmosphere and have a reduced leaf area.

The water flow in the stem, D, is in principle equal to the flow that transpires from the leaves, T, but there are differences regulated particularly in conditions of intermittent light. If the light decreases abruptly, T decreases instantly, while D remains constant. This can be studied with a double lysimeter.

The water flow may be partly interrupted in particular conditions, other than the closure of stomata or soil dryness. The soil may be frozen, the tree may be twisted because of strong wind, or the trunk may undergo alternative freezing and thawing. In all these cases the water column in the vessels is broken and air bubbles block the circulation: there is embolism. Trees with wide vessels are particularly sensitive to this in winter and early spring (Milburn, 1966; Wang et al., 1992). Conifers are clearly more resistant.

1.4. Water that is lost (the great leap and its regulation)

A body of water loses a quantity of vapour proportionate to the gradient established between the immediate surface (where there is saturation) and a certain distance (boundary layer) in the dry air, where the water vapour content is lower. The gradient is accentuated, through the increase of $\Delta\Psi$, if the water surface is warmer than the surrounding air, or through decrease in the gradient in space when the boundary layer reduces when there is wind. The values of this potential evaporation range from 10 to 15 mm of water lost per day (or 10 to 15 kg/m^2) in desert or subtropical semi-desert, to around 2 mm on an average for a temperate oceanic environment during the vegetation period. In comparison, the real potential evaporation of a humid soil is always lower because of resistance to water circulation in the soil. The leaf area also has an impact. An essential parameter is the gradient of water vapour between the air outside and the interior of the plant, which depends closely on the temperature (Fig. 3.2).

1.4.1. The physical aspect of transpiration

The transpiration of plants follows the law of evaporation of liquid surfaces, with modifications that are due to conductance of water vapour in structures existing between the sources of liquid water (cells) and the ambient air. There is cuticular transpiration, due to the passage of vapour across the cuticle, often low, because the conductance of the cuticle is very low. There is cortical transpiration, also quite low because the suberized tissues of the stem and branches are quite thick. Above all, there is stomatal transpiration, which can be regulated. Beyond the leaf, the

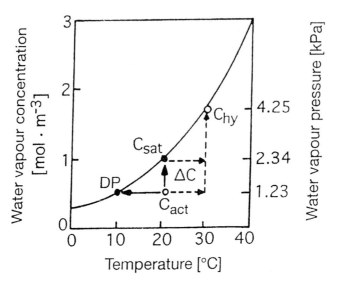

There is often a steep vapour gradient between the interior of a leaf (C_{hy}) and the outside air (C_{act}) for two reasons: Very often, $C_{act} < C_{sat}$, the concentration when the outside air is saturated with water. Moreover, during the day the leaf receives light. It absorbs energy partly converted into heat, which makes the leaf warmer than the ambient air, and causes the gradient to be $\Delta C = C_{hy} - C_{act}$. C_{hy} is in fact the concentration at saturation for the internal temperature of the leaf. If the temperature of the air decreases but remains steady, which happens during a calm night, the decrease may cause relative humidity to reach 100% when it reaches DP, the dew point. Fine drops of water will be deposited on the leaf. The form of the curve shows also that in a hot climate the water losses can be much greater than in a temperate climate, even if the irradiation is not very different.

Fig. 3.2. The maximal water vapour content of the atmosphere at different temperatures (*Larcher, 1995*).

boundary layer, due to the accumulation of water vapour in the air in contact with the leaf (Fig. 3.3) at a value close to saturation, brings a final resistance. In the case of thallophytes, mosses, and marine grasses that can be greatly exundated, these distinctions have no purpose and we can go straight to the overall transpiration, which depends on the exposed surface of the plant and the more or less extended form of the thallus and the boundary layer.

For each species, in the normal conditions of its habitat, stomatal transpiration may reach a value considered maximal, which happens when the stomata are fully open, the resistance of the boundary layer is low, and there is nothing particular to prevent water from reaching the stomata. Great differences can be observed between plants (Tables 3.5 and 3.6) according to ecological characters, biological forms (e.g., plants with persistent leaves as opposed to caducous leaves), and overall morphology (erect, prostrate, or tuft shape). Very high values are found, for example, in erect grasses of marshlands (helophytes), or megaphorbiae. Low values

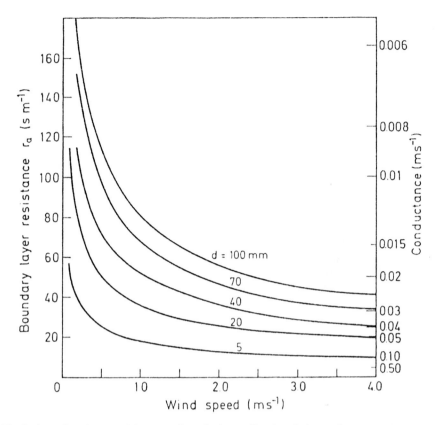

Y-axis, boundary layer resistance and conductance. X-axis, wind speed.

Fig. 3.3. Variation of resistance of boundary layer
(Grace, 1977).

are found in trees of the rainforest, but this is because the air is naturally very humid.

Water escapes the leaves by diffusion but also en masse from sub-stomatal chambers where, in sunlight, a mass of hot air saturated with moisture escapes by the simple fact of its suppression induced by heating (PV = nRT). The conductance of water vapour, ^CH_2O, is the inverse of the resistance to diffusion, or "resistance to transpiration".

Rate of release: $^VH_2O = {}^CH_2O \cdot \Delta[H_2O]$. $[H_2O]$ is expressed in g/m³ or mol/m³.

The epidermal resistance (R_{ep}) to transpiration results from stomatal resistance (R_s) and cuticular resistance (R_{cut}). The epidermal conductance (C_{ep}) is equal to $1/R_{ep}$.

$1/R_{ep} = 1/R_s + 1/R_{cut}$, and conductance can also be expressed in cm/s or in mmol/m²/s.

Table 3.5. Maximal rates of leaf transpiration, or stem transpiration (succulent plants) in typical conditions of the habitats (data from several authors, in Larcher, 1995)

Plant type	Rate (μmol $H_2O/m^2/s$)
Tropical humid zone	
Trees of rain forests	to 1800
Trees of *nebelwald* (mist forests)	400, sometimes 2000 to 3000
Lianes	to 2000
Tropical zone in dry season	
Palm trees	1200 to 1800
Mangroves	600 to 1800
Trees of dry forest	800 to 1400
Subtropical deserts	
Shrubs	2800 to 7000
Desert plants	1000 to 5000
Mediterranean sclerophyte trees and shrubs	600 to 4000
Temperate or cold zones	
Deciduous trees adapted to sunlight	2500 to 3700
Deciduous trees adapted to shade	1200 to 2200
Conifers	1400 to 1700
Chamaephytes of the tundra region	150 to 450
Alpine chamaephytes	1800 to 3000
Herbaceous dicotyledons	
some megaphorbiaceae	9000 to 11,000, sometimes 16,000
heliophytes	5200 to 7500
sciaphytes	1500 to 3000
Monocotyledons	
of tundra	200 to 350
of prairie	3000 to 4500
of temperate humid zones	5000 to 10,000
of dunes	2000 to 4000
halophytes	1200 to 2500, sometimes 4500
aquatic plants	5000 to 12,000
Succulent plants	
with developed leaves	800 to 1800
with succulent stems	600 to 1800

$C_{ep} = C_s + C_{cut}$. C_{cut} is fixed, and C_s varies (Fig. 3.4), under the effect of internal and external factors.

C_{ep} increases linearly with the width of stomatal opening; the maximum conductance depends not only on the size of the possible opening, but also on the shape, position in relation to other cells, and density of stomata.

Stomatal conductance does not vary much according to the species (Korner, 1994), ranging from 100 mmol/m^2/s for certain fleshy plants to 500 mmol/m^2/s for some cultivated heliophytes.

Cortical transpiration of a tree represents only around 1% of the total transpiration.

Table 3.6. Daily transpiration from plants in average conditions of their desert habitats, in wet season and dry season (Stocker, 1970, 1974; Caldwell et al., 1977; Nobel, 1977, in Larcher, 1995)

Plant	Wet season	Dry season	Residual transpiration (%)
Saharan or sub-Saharan plants			
Nitraria retusa	210	165	78
Zylla spinosa	240	150	62
Zygophyllum coccineum	165	80	48
Pennisetum dichotomum	165	65	39
Hammada scoparia	(4)	(1.5)	38
Haloxylon persicum	280	100	36
Anabasis articulata	(3.1)	(1.0)	32
Retama retam	270	80	29
Artemisia herba-alba	(6)	(1.6)	27
Noea mucronata	(5.5)	(1.0)	18
Halophytes of Utah			
Atriplex confertifolia (C4)	155	30	19
Ceratoides lanata (C3)	154	2.2	1.4
Cactus of California			
Ferocactus acanthoides	17	0.35	2

The values are in mol $H_2O/m^2/d$ or sometimes (parentheses) in gH_2O/g fresh matter/d.

Cuticular transpiration may vary the most according to the ecology of the species. It corresponds to diffusion across a thickness of hydrophobic medium (occupied by wax and cutin) and is low to the extent that the cuticle is thick. It commonly varies from 20 $mmol/m^2/s$ in some cultivated plants or sciaphytes to just 0.5 $mmol/m^2/s$ in desert succulents (Korner, 1995).

1.4.2. Physiological control of transpiration

Thallophytes must be distinguished from the higher plants. The thallophytes have a physiological activity that is closely dependent on the activity of the external water, which very quickly reaches equilibrium with the activity of cell water (Fig. 3.5).

Transpiration is strictly dependent on physical factors only to the extent that stomatal conductance remains constant. If the physical factors tend to vary so as to considerably increase the transpiration, the plant may have to decrease it, and it can do that by reducing its stomatal conductivity. Conversely, a plant that suffers desiccation and closes its stomata (abscisic acid being a mediator: Fig. 3.4) benefits from opening them during a humid period, that is, to considerably increase conductance, which will allow it to resume photosynthetic activity. The capacity of plants to regulate stomatal opening allows them to maintain their hydric equilibrium in a changing environment. In relation to the potential evaporation, the plant, depending on its needs, may reduce transpiration

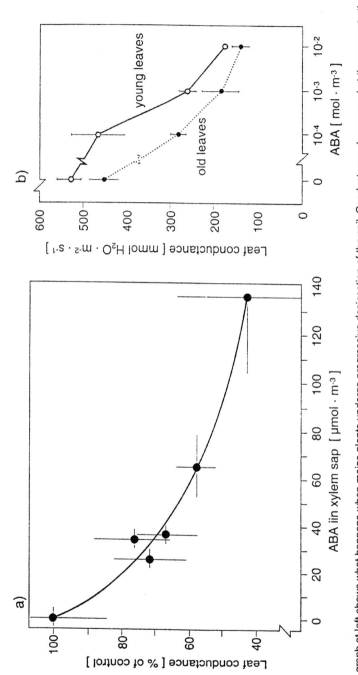

Fig. 3.4. Stomatal conductance and the presence of abscisic acid
(*Zhang and Davies, 1990; Henson and Turner, 1991, in Larcher, 1995*).

The graph at left shows what happens when maize plants undergo progressive desiccation of the soil. Conductance decreases, but the concentration of abscisic acid (ABA) observed in the sap increases. The graph at right shows the effect of the application of known solutions of ABA in leaves of *Lupin luteus*. The depressive effect of transpiration by ABA seems even more marked in young leaves than in the old ones.

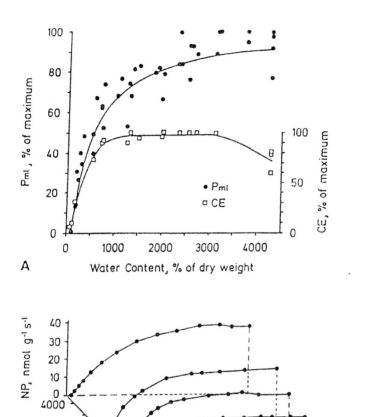

A

B

Photosynthesis was measured at 15°C (tundra plant) and under conditions of saturated light (200 μmol/m²/s) but not too much light in order to avoid photoinhibition, and at increasing CO_2 levels (Ca, curve B). Very moist sphagnum (around 40 masses of water for a mass of dry matter) was separated for a preliminary measurement as a function of the CO_2 concentration. It was allowed to dry naturally during the days following and a new series of measurements were taken every day. From the original B curves, the A curves were calculated. Photosynthesis NP and P_{ml} increase greatly with hydration, but it is seen that the slope of increase with the CO_2 level (CE, curve A) reaches a maximum at average hydration. Thus, when hydration is very high, the CO_2 efficiency seems to decrease, perhaps because of problems of diffusion.

Fig. 3.5. Influence of hydric state of sphagnum on photosynthesis and carboxylation efficiency *(Tenhunen et al., 1995, in Schulze and Caldwell, 1995).*

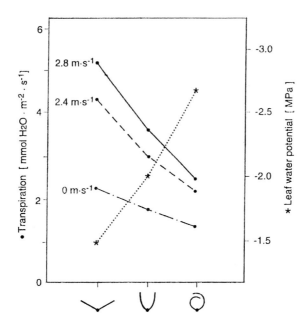

The study was conducted on rice (*Oryza sativa*). The leaf rolls when its water potential decreases. Although this does not enable it to decrease its transpiration much when the air is calm, the economy achieved is spectacular when there is wind.

Fig. 3.6. Adaptation to drought by rolling of leaves in monocotyledons
(*O'Toole et al., 1979; Hsiao et al., 1984, in Larcher, 1995*).

more or less, and not only by means of stomata (Fig. 3.6). It may do so by a passive strategy ("automatic" closure, if the plant tends to lose too much water) or active regulation. The plant reacts more particularly to dry air, wind, toxic gases, and thermal aggressions. The basic mechanisms of regulation rely on the thresholds, rapidity, and efficiency of its reaction.

A low threshold of $\Delta\Psi H_2O$ and rapid closure often go together. A spring plant in the undergrowth such as *Alliaria officinalis* can close its stomata in 2 to 4 minutes. On the other hand, a Mediterranean heliophyte such as geranium (*Pelargonium zonale*) takes half an hour to close its stomata; thus it loses a great deal of water. The geranium, however, is more efficient. A cut leaf of *Alliaria* dries out in 10 to 20 min. despite the closure of stomata; a cut leaf of geranium remains hydrated and supple for several hours. The difference arises from the high cuticular resistance of geranium. Differences can be observed in the same species depending on the ecotypes of simply phenotypic variants, and within a single individual between leaves exposed to sun and those that remain in the shade.

The efficiency of reaction can be estimated from the ratio of transpiration with open stomata to transpiration with closed stomata. For *Alliaria* the ratio is about 3, for plants of the semi-desert a ratio of about 20 to over 100 is observed, in natural conditions. For a comparison between varieties, or more strictly between species, we can calculate the ratio of conductances C_{cut}/C_{st}, which may be from about 0.1 to 0.3 for grasses of temperate oceanic climates to just 0.01 to 0.005 for succulents.

1.5. Hydric equilibrium of the entire plant

1.5.1. A dynamic equilibrium

The activity of transpiration is strongly influenced by the CO_2 content of air (Fig. 3.7). In this context, the effect of high CO_2 levels has been thoroughly studied, to test in particular the beneficial effects (Table 3.7) on water use efficiency. The effective CO_2 content of the leaf surface depends on the boundary layer, which varies as a function of the wind speed, and on light intensity and its fluctuations.

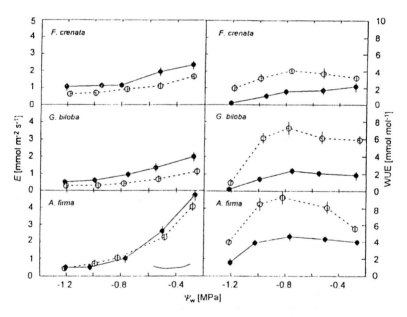

Three-year-old trees of Asiatic temperate forest were transplanted in open rooms 5 months before the measurements were taken. The species studied were *Fagus crenata*, *Gingko biloba*, and *Alnus firma*. WUE, water use efficiency, mmol net fixed CO_2/mol water evaporated under 1200 µm/m²/s PPFD and at 25°C ± 1.5. E, rate of evaporation, mmol/m²/s. Closed circles, plants in chambers at 700 µmol/mol CO_2. Open circles, plants in chambers at 350 µmol/m CO_2. The water potential of soil is given in MPa.

Fig. 3.7. Interactions between water potential and high levels of CO_2 measured on transpiration and water use efficiency of some deciduous trees
(*Liang et al., 1995*).

Table 3.7. Water use efficiency measured over long periods during vegetative growth
(Polster, 1967; Simpson, 1981, in Larcher, 1995)

	Water use efficiency (g dry matter/kg of water lost)
Sunflower (*Helianthus annuus*)	1.7
Bean (*Phaseolus vulgaris*)	1.9
Wheat (*Triticum aestivum*)	2.3
Oak (*Quercus sp.*)	2.9
Pine (*Pinus sylvestris*)	4.3
Beech (*Fagus sylvatica*)	5.9
Maize (*Zea mays*)	4.0
Cactus (*Opuntia* sp., CAM)	10.0

The deviation in relation to the equilibrium can be studied over various time intervals, from tens of minutes, in fluctuating light, to a 24-h day, on the seasonal scale. The deviation in relation to an "ideal

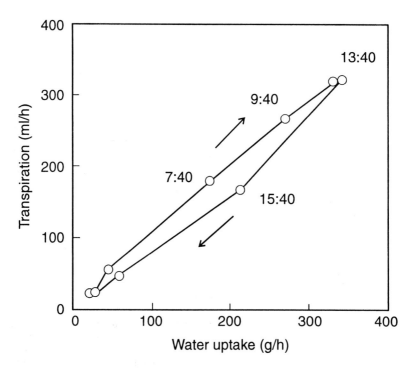

Measurements were taken on a sunny day, from four tomato plants, every two hours. A phenomenon of hysteresis was observed, with transpiration higher than absorption in the morning and the reverse in the evening.

Fig. 3.8. Simultaneous measurement of transpiration and absorption of water
(Andriolo et al., 1996).

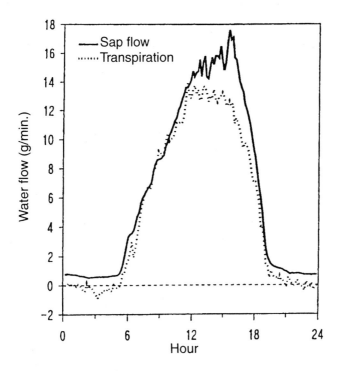

Measurement of sap flow by assessment of heat is precise and quick, allowing us to follow fluctuations of a few minutes. During the night the sap flow is greater than transpiration. During the day there is a slowing of sap flow with respect to transpiration, which can be explained by the use of water reserves the plant has in the morning. The sap flow curve is consistently above that of transpiration, which indicates that the measurements of transpiration by gaseous exchanges seem systematically underestimated. Corrections in atmospheric pressure have further helped the authors to improve the precision of measurements.

Fig. 3.9. Measurement of sap flow and transpiration
(Ameglio et al., 1993).

equilibrium" is, independent of the water serving to increase the biomass, negative if the water is released faster than it enters, or positive if the water that enters is greater. On the basis of 24 h, the deviation is negative during the day and positive during the night. On the basis of daily fluctuations of light, phase difference is observed (Figs. 3.8 and 3.9).

The situation must also be considered in comparison of different plant organs. Leaves in sunlight may lose a great deal of water and, independent of what is drawn up by the roots, draw water from organs that have the equivalent of water reserves, for example, the parenchyma of the stem or the pith. This is what leads to a reduction in the trunk diameter of a tree during a sunny afternoon. Soft fruits, which often

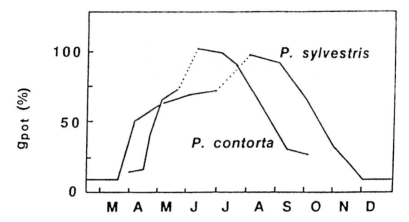

Pinus sylvestris, found in the Tyrol, faces m...der climatic conditions and transpires for a longer period than *P. contorta*, which is found in Montana. The daily maximum value (g_{pot}) is highest in that season in which the formation of new needles ends (dotted portion of line). This daily maximum corresponds to the maximum possible value of transpiration during the course of the year (g_{max} in the original study).

Fig. 3.10. Daily maximum rate of transpiration as a function of the season, in conifers
(Korner, in Schulze and Caldwell, 1995).

contain 90 to 95% water, may also serve as a reserve. INRA has developed a system of automatic alert of lack of water in lemon trees by measuring the reduction of lemon diameter, which occurs as soon as the plant feels the deficiency of water.

Transpiration varies not only over a day but also as a function of the seasons in relation with the climate and its more or less rigorous constraints (Fig. 3.10).

Also, on the scale of an ecosystem, great variations are observed according to the spatial position of individuals, for example in a shady tropical forest (Fig. 3.11), which brings us to the dynamics of the entire system.

1.5.2. How can a negative drift be avoided?

It is relatively convenient to distinguish plants of a humid environment and those of an arid environment. For the former, there is often a root system and a conducting system that can only bring a small quantity of water in comparison with extreme needs. Thus, often a "midday" depression of transpiration is observed that extends from the late morning to at least the middle of the afternoon. If there is a period of several consecutive dry days, the plant may go from a situation in which there is no regulation to one in which there is stomatal closure in the middle of the day, then, the soil becoming progressively drier, to a situation of general regulation all through the day, with likely a permanent closure of stomata.

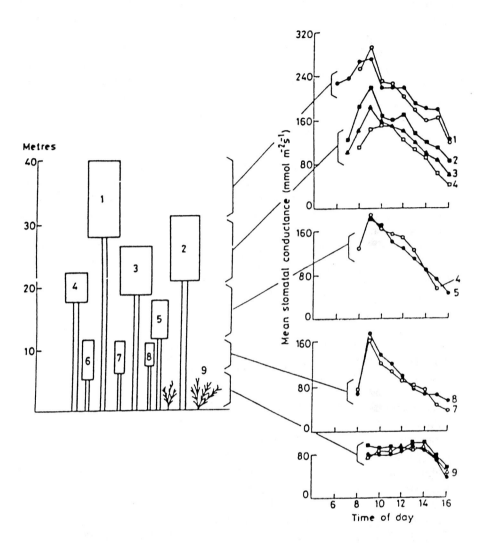

Leaves of the tallest trees have the greatest conductance (and thus transpiration per unit of area). Conductance decreases slightly for the medium trees (no. 2 to 8), but keeps the same pace, with a clear maximum at about 9.00 h and a progressive decrease up to 16.00 h. Shrubs of the undergrowth (no. 9) show a different behaviour. If the conductance is low enough (it is also linked to the low photosynthetic activity), it decreases only slowly in the afternoon, because these plants are in a much more protected environment, and moreover lose little water.

Fig. 3.11. Daily variation in stomatal conductance as a function of the location in an Amazonian forest *(Dolman et al., 1991, in Schulze and Caldwell, 1995).*

Plants of arid regions generally have a relatively reduced leaf area in relation to the size of the trunk (conduction) and the size of the root system (absorption), which goes very deep into pockets of water or the water table and/or extends widely under the soil surface. The plant will thus show optimal transpiration, as far as the usable water reserves (internal and external) are sufficient. If the drought lasts a few months, however, regulatory mechanisms will intervene. In a Mediterranean plant such as green oak, for example, a reaction to general summer dryness is observed, in the form of reduction of stomatal conductance (the stomata open only slightly) and also reduction of the period of day during which the stomata open.

The physiology of stomatal closures relies on sometimes passive and sometimes directly active phenomena. If the leaf tends to lack water, all the stomatal cells and stomata close, passively. Photosynthesis will also intervene, more specifically: as soon as the ΨH_2O decreases a little too much, photosynthesis diminishes considerably. Less CO_2 is used, and it is thus more concentrated and causes the closure of stomata. Direct alert systems may quickly intervene, which will allow the plant to avoid "foolishly" losing water. In willow, the intervention of electrophysiological signals and hormonal signals is observed. In many species in which this has been studied, from the time the soil dries out a little, the roots emit abscisic acid, which migrates rapidly to the leaves following the flow of sap and causes the closure of stomata, which can be observed in the field or by experimental application of ABA (Fig. 3.12) and leads to a slowing of growth.

1.5.3. General comparison of hydric states

Direct field measurements of water that enters and water that leaves the plant are not easy. But the water content of leaves of a species or variety at a given moment can easily be measured. This can be compared to the water content of leaves in plants of the same variety, and at the same stage of development, well supplied with water and fully turgescent. By this means we obtain a quantity of water at a saturation Q_{sat}, which can be compared to quantities of water in the leaves observed (Stocker, 1929; Ochi, 1962) in the field Q_{real}. The relative water content is calculated as follows:

$$RWC = Q_{real}/Q_{sat} \times 100 \ (\%)$$

The measurement of relative water saturation deficit is more informative:

$$WSD = (Q_{sat} - Q_{real}/Q_{sat}) \times 100 \ (\%)$$

Another method is to observe the variations in osmotic potential $\Delta \Psi p$.

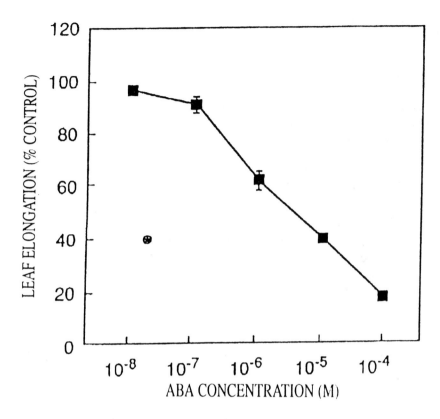

The above-ground part is detached and soaked in solutions. The squares show the increase in length of emerging leaves, after 8 to 16 h of treatment. The isolated circle shows the relation between the activity (on growth) of the sap isolated from drought-stressed plants in the field and its effective ABA concentration (1.4×10^{-8} M).

Fig. 3.12. Leaf elongation in wheat seedlings as a function of ABA concentration
(Munns, 1992, in Munns and Sharp, 1993).

$\Delta\Psi p$ (equivalent to $\Delta\Psi H_2O$) varies very little when the cells remain turgescent (little variation in volume) but when the cells enter into plasmolysis Ψp increases a great deal. However, this is not necessarily representative of the water content because the plant, particularly in the case of halophytes, can synthesize osmoregulator metabolites such as proline and certain sugars (e.g., inositol) or absorb certain ions, in order to keep a high water content at the cost of an increase in Ψp.

A more sensitive and direct measurement of changes (Ramball, 1992) is that of leaf water potential, which can easily be measured in a pressure chamber.

2. MINERAL SUPPLY

2.1. Introduction

Mineral supply is based on two sources, soil water, in which the activity of soil bacteria provides nitrate, and elements found in the soil, which intervene mostly as a source of cations by means of their cation exchange capacity (Table 3.8). Also, rainfall not only brings water (Table 3.9) but also is an important source of mineral elements, particularly in the temperate zones. The absorption of minerals from soil may be facilitated by high water potential, or it may be made difficult if the potential is low, even if there is apparently a large quantity of water.

Table 3.8. General properties of soil ion exchangers
(Jeffrey, 1987; Kuntze et al., 1988, in Larcher, 1995)

Exchanger	Specific area of exchange (m^2/g)	Cation exchange capacity (eq/kg)
Hydrated oxides of Fe or Al	25-40	0.03-0.05
Kaolinite	10-20	0.03-0.15
Illites	100-300	0.2-0.5
Vermiculite	600-800	1-2 (8)
Montmorillonites	700-1200	0.8-1.2 (10)
Humic substances	800	1.5-5

The values in parentheses are exceptional. Kaolinite, vermiculite, montmorillonites, and illites are derived from clays.

Table 3.9. Uptake of minerals from rainwater $(g/m^2/year)$
(Golley et al., 1975; Kallio and Veum, 1975; Likens et al., 1977)

Element	Tropical rain forest (Central America)	Temperate deciduous forest (N. America)	Tundra (Scandinavia)
N		2.07	0.07 to 0.1
S		1.88	0.5 to 0.6
P	0.10	0.0004	
Ca	2.93	0.22	0.25 to 0.54
Mg	0.49	0.06	0.05 to 0.15
K	0.95	0.09	0.08 to 0.12
Na	3.07	0.16	0.1 to 0.4
Fe	0.30		

2.2. Absorption of mineral ions by plant cells

Mineral ions are generally absorbed mainly by cells of the rhizoderm (and their extensions, the absorbent hairs). Cells of the rhizoderm have a very contorted plasmalemma, which increases the area of exchange with the exterior. The plasmalemma has a membrane charge of about –120 mV (interior to exterior); the positive charge is on the exterior. This potential, like the presence of an often acid exterior pH, and a very dilute exterior medium, leads to a certain number of consequences for membrane transport (Fig. 3.1) that are either spontaneous or active. The vacuole serves as a transitory sink.

Some ion channels realize a uniport for a cation and others for an anion. In view of the membrane charge, the cations enter and the anions leave, and an equilibrium must be established between entries and exits of charges (with respect to membrane potential). Hydrolysis of ATP by membrane ATPases allows the expulsion of protons to the outside. These protons, by soil acidification, facilitate the liberation of useful cations (Fe^{+++}, Ca^{++}, K^+, etc.) and they also allow the symports and antiports, symports H^+ and anions, antiports H^+ and cations, respecting the equilibrium of the charge and allowing the absorption of nitrate and evacuation of excess cations (Na^+, for example). These cotransports use the energy of the gradient of protons constructed by ATPase activity. Specific transporters exist at the membrane level. For example, in *Arabidopsis thaliana*, clones of which can live in soils poor in soluble iron, a highly effective chelator has been indicated (Piglivcci et al., 1995).

At the level of the vacuole, tonoplastic ATPase allows evacuation of the protons in the vacuolar liquid, and an antiport H^+/Ca^{++} has been observed, which is regulated by calmodulin. There are also transporters with malate or nitrate, and other antiports with cations. The vacuolar accumulation of K^+ and other ions is essential to the mineral supply to the plant and the drop in water potential, which will enable absorption of water.

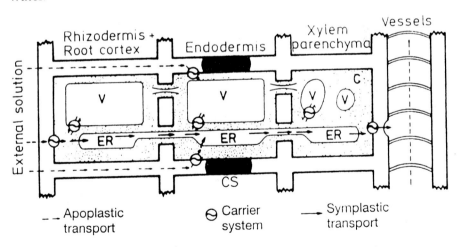

V, vacuole. ER, endoplasmic reticulum. CS, Caspary framework. C, cytoplasm. Small arrows, apoplastic transport. Large arrows, symplastic transport.

Fig. 3.13. Symplastic and apoplastic routes
(*Luttge, 1973; Lauchli, 1976, cited in Larcher, 1995*).

2.3. Absorption by roots

Absorption by roots involves not only the cells of the rhizoderm. Water in particular may follow the cell walls to migrate to the endoderm, which is called **apoplastic circulation**. Apoplastic circulation occurs passively, according to a gradient, and nothing regulates it directly. It occurs in what is called the apparent free space. In the Caspary frameworks (dicotyledons) or U cells (monocotyledons), all that circulates by the apoplastic route (and that is in continuity with the exterior field) must rejoin the cell interior (Fig. 3.13) and the flow that comes from cells of the rhizoderm. The circulation that occurs in the cells is called **symplastic circulation**. It is most likely produced by the endoplasmic reticulum, which, like hyaloplasmic gel, is continuous from cell to cell by the plasmodesmata. The vacuoles constitute intermediate reservoirs, but the essence of mineral substances, brought to the cells of the parenchyma from the xylem, leaves it by means of membrane transporters, to again reach the apoplastic domain and rise in the plant by the vessels or even the tracheids.

Water circulates rapidly in the xylem, carrying all the minerals absorbed by the roots. The minerals travel essentially towards the growing organs, through the xylem as well as the phloem, the communication between the two types of conducting tissues being possible at the level of roots and also at the level of stem nodes. Ions, however, circulate essentially in the xylem (Fig. 3.14). The phloem is mostly used when an organ (e.g., a leaf) stops growing, and in that case it can, for example,

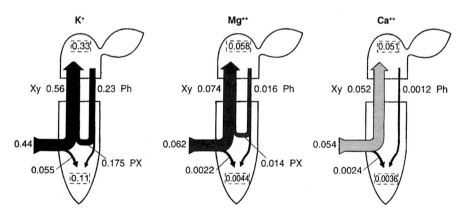

Measurements on lupin (*Lupinus albus*). The boxed figures indicate the levels in above-ground and underground parts, the others the flows in μmol/g fresh matter/h. PX, flow from phloem to xylem in the roots. XY, xylem. Ph, transport by phloem.

Fig. 3.14. Transport and remobilization of K⁺, Mg⁺⁺, and Ca⁺⁺ between the root and the above-ground part (*Jeschke et al., 1985, in Larcher, 1995*).

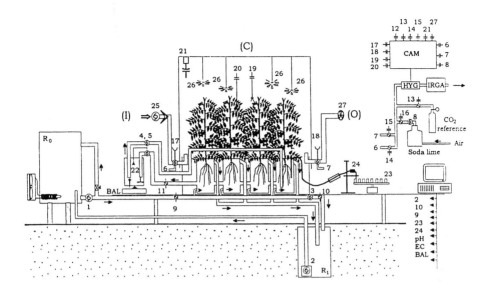

Tomato plants in hydroponic culture were studied. The set-up was located in a greenhouse simulating conditions of commercial cultivation. The experimental chamber contained a row of 4 tomato plants (length 1.40 m) and measured 1.60 m length × 1.40 m height × 0.60 m depth. The air was homogenized by ventilators (26). A pump (25) brought in air, and an anemometer (27) measured the flow of air at the outlet.

With pumps (6, 7, 8) and solenoid valves (13 to 16) controlled by computer (CAM), the standard gases or the gases at the entrance and exit of the chamber could be sampled alternately. CO_2 concentrations were analysed by IRGA, which received the standard gases at the beginning of the experiment. Water vapour was measured using a hygrometer (HYG) with a mirror at dew point, which enabled measurement of transpiration. Air- free of water and CO_2 was obtained with soda lime.

A quantum sensor (21) measured the light in the chamber. Various temperature captors were required—17 and 18 at the entrance and exit of the chamber, and 19 and 20 on the lower surface and upper surface of leaves.

The supply of water and minerals was controlled. The nutrient solution R_0 was supplied by pump (1) and then by controlled opening of the valve (9). The solution returned by pump (2) towards R_0. A pump (3) and a solenoid valve (10) provided drainage at the end of the experiment.

A pot without a plant was placed on a balance outside (22 and BAL) but connected to the chamber. Two pumps (4 and 5) controlled by a water level probe (22) constantly circulated the solution between the pot on the balance and the hydroponic system. The nutrient solution was renewed every 3 h. The solution was measured before and after the experiment from samples of 20 ml collected by means of a peristaltic pump (24) on an automatic collector (23). This system allowed a simultaneous measurement of the quantity of water lost by the system and the absorption of minerals.

Fig. 3.15. A set-up for simultaneous study of photosynthesis, absorption of water, and absorption of nitrate
(Andriolo et al., 1996).

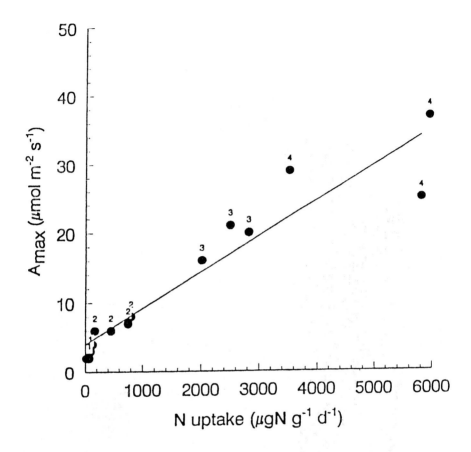

Many observations show that photosynthetic production is favoured by the nitrogen-richness of the soil. The influence of the effective nitrogen absorption rate can be probed and studied. A_{max}, maximal rate of photosynthesis in optimal natural conditions of humidity, temperature, and light. N uptake, daily nitrogen absorption rate. The relation is well-described by the regression line, r = 0.940. Four groups of species were studied in relation with their mycorrhizal status. They are listed in order of increasing productivity:

1. *Vaccinium vitis-idea, Calluna vulgaris, V. myrtillus,* and *Rhododendron ponticum,* which have ericoid mycorrhizae.
2. *Betula pendula, Fagus sylvatica, Quercus robur,* and *Acer pseudoplatanus,* which have ectomycorrhizae.
3. *Eupatorium cannabinum, Potentilla reptans,* and *Verbena officinalis,* which have vesicular arbuscular mycorrhizae.
4. *Senecio vulgaris, Lolium perenne,* and *Triticum aestivum,* which do not have mycorrhizae.

Fig. 3.16. Action of nitrogen uptake on maximal photosynthesis
(Schulze and Caldwell, 1995).

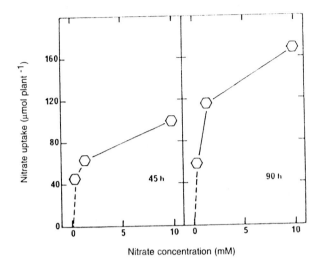

Nitrate is absorbed as a function of the concentration according to a curve that seems to be of the Michaelis type. It is also possible to note that in plants that have had 90 h of nitrate at their disposal absorption is more rapid than when they have had 45 h, perhaps because of the better general state of the plant. The study was done on 2-month-old seedlings of peach tree.

Fig. 3.17. Nitrogen uptake in a young tree that was previously deficient
(Gojon et al., 1991).

reexport part of its K^+ and Mg^{++}. The alkaline earths, however, are not remobilized much, especially Ca^{++}, so the Ca/K ratio of leaves increases with age. The ions remobilized from old leaves are directed towards the young leaves or, particularly in the case of perennial or ligneous plants, towards reserve organs (underground parts and trunk) at the end of summer and in autumn.

Absorption of minerals is greatly influenced by photosynthesis, particularly absorption of nitrates (Fig. 3.15). Inversely, nitrogen supply affects photosynthetic activity (Fig. 3.16). Absorption varies also as a function of previous deficiencies (Fig. 3.17).

2.4. Mineral content and use of minerals by plants

The mineral content (Table 3.10) reflects the needs of the plant, in which minerals are distributed according to the source-sink relationships between the different organs (Table 3.11), and the composition of the nutrient soil. Many plants of an acid medium, for example, contain aluminium, which is not strictly useful, and can even prove harmful to certain species (Table 3.12) (see also Chapter 5).

Many cabbages are very rich in sulphur, which corresponds to the peculiarity of their metabolism. Certain minerals pass into tissues after

Table 3.10. Mineral content of different types of vegetation
(Rodin and Bazilevitch, 1967, in Larcher, 1995)

Biotope type	Ash characteristics	Overall mineral content	Rate of litter decomposition	Type of vegetation
Nitro-boreal	N > (K, Mn)	Low	Slow	Tundra
	N > Ca	Low	Slow	Taiga
	N > Ca (Si, Mg)	Medium	Slow	Mixed forest
Nitro-arid	N > Ca (Na, Cl)	Medium	Very rapid	Deserts with shrubs
Nitro-subtropical	N > Ca (Si, Al, Fe)	Medium	Rapid	Deciduous forests
Calco-temperate	Ca > N	Medium	Moderate and highly humiferous	Oak or beech forests
Calco-subtropical	Ca > Si (Al, Fe)	Medium	Very rapid	Sahelian vegetation
Silico-semi-arid	Si > N	Medium	Very rapid	More or less high steppes
Silico-arid	Si > N (Na, Cl)	Medium	Very rapid	Semi-ligneous, desert therophytes
Silico-tropical	Si > N (Fe, Al)	Medium	Very rapid	Savannahs
	Si > N (Al, Fe, Mn, S)	Medium	Very rapid	Tropical rain forests
Saline	Cl > N	High	Very rapid	Halophytic vegetation

Table 3.11. Distribution of total reduced nitrogen between above-ground and
underground parts (Gojon et al., 1991)

Nitrate concentration in nutrient medium (mM)	Total reduced nitrogen/plant		
	Roots	Stems	Whole plant
0.5	255 ± 35	1030 ± 60	1290 ± 80
1.5	237 ± 26	1190 ± 110	1420 ± 120
10	256 ± 58	1160 ± 110	1410 ± 140

In peach seedlings, nitrogen content (except NO_3) practically did not change with nitrate content in the medium; 0.5 mM was sufficient for the needs of the plant. It must also be noted that the above-ground parts were much richer in nitrogen than the underground parts, even when the biomass was very similar.

reduction at the level of organic matter (see Chapter 2), particularly nitrogen and sulphur, which are mainly absorbed in the form of nitrate and sulphate.

Certain taxonomic groups, or even certain types of environments, may have peculiarities. Families such as Graminaceae, Cyperaceae, as well as Juncaceae are always very rich in silicon, which may often represent more than half of the total mineral content. Nitrophilic plants

Table 3.12. Species (mostly calcicolous) that are particularly sensitive to aluminium in acid soils (Grime and Hodgson, 1969, in Larcher, 1995)

Creeping bent	*Agrostis stolonifera*
Beetroot	*Beta vulgaris*
Wheat (many cultivars)	*Triticum aestivum*
Meadow fescue	*Festuca pratensis*
Lettuce	*Lactuca sativa*
Lucerne	*Medicago sativa*
Barley	*Hordeum vulgare*
Pincushion flower	*Scabiosa columbaria*

have more nitrogen than potassium, and generally a great deal of nitrogen is found in plants growing in nitrogen-rich soil, such as soils found in many riverside forests, pastures, and ruderal zones. Na^+, Cl^-, and SO_4^{--} are abundant in halophytes. Silicon is found in large quantity in tropical plants, because forest soils contain a great deal of it and also because high temperatures favour the solubility of SiO_2. Calcium is abundant not only in plants of chalky land but also in those of arid subtropical regions.

Plants of quarry areas, or close to non-ferrous metal factories, may be very rich in certain heavy metals (see Chapter 5). All this means that certain wild plants can be used as indicators of the nutrient value of the soil, or even to detect the presence of mineral deposits.

2.5. Mineral quality of soils and plant distribution

Finally, many plants have complex relationships with soils, where the mechanical aspect and chemical aspect interfere in conjunction with the physiological diversity of species, or groups of species, either taxonomic or biological. Calcicolous plants, for example, are distinguished from calcifugous plants. At the taxonomic level (Table 3.13), we find vicariant species, i.e., species very close in botanical terms, probably having a common ancestor, of which one is calcicolous and the other is calcifugous.

In other words, the nitrogen uptake depends a great deal on the climate and soil quality (Table 3.14). Rates of nitrogen mineralization vary greatly depending on the biomes, by means of soil microflora.

3. RESULTS: DIVERSITY OF POLICIES FOR WATER MANAGEMENT DEPENDING ON SPECIES AND HABITATS

3.1. Introduction

Certain plants appear hydrostable and show only slightly varying water potential, in principle not falling to very low values. This is the case with aquatic plants, plants of the undergrowth, and many trees and plants of oceanic temperate regions. Other plants are euryhydric and tolerate great

Table 3.13. Examples of vicariants in the Alps, on the basis of interaction with calcium or on the basis of geography (Favarger and Robert, 1962; Landolt, 1971; Ellenberg, 1986; Klein, 1992)

Genus	Calcicolous species	Calcifugous species
Achillea	*atrata*	*moschata*
Androsace	*helvetica*	*alpina*
Carex	*firma*	*curvula*
Cerastium	*latifolium*	*uniflorum*
Doronicum	*grandiflorum*	*clusii*
Gentiana	*clusii*	*kochiana*
Gentiana	*lutea*	*punctata*
Hutchinsia	*alpina*	*brevicaulis*
Leontodon	*montanus*	*helveticus*
Primula	*auricula*	*hirsuta*
Pulsatilla	*alpina*	*sulphurea*
Ranunculus	*alpestris*	*glacialis*
Rhododendron	*hirsutum*	*ferrugineum*
Saxifraga	*moschata*	*exarata*
Soldanella	*alpina*	*pusilla*
	Western Alps	Central Alps
Senecio	*incanus*	*carniolicus*
Hieracium	*peletierianum*	*hoppeanum*

Table 3.14. Rates of nitrogen mineralization in upper soil horizons, from litter decomposition during the vegetation season (Larcher, 1995, on the basis of data from numerous authors)

Plant formation	Net mineralization rate, mg N/m^2
Tropical zone	
Rain forest	10 to 80
Deciduous forest in dry season	10 to 20
Gallery forest	7
Savannahs	0.3 to 0.5
Itinerant agriculture	7 to 10
Temperate zone	
Mediterranean sclerophyte forest	to be studied
Conifer forest	3 to 12
Deciduous forest	10 to 25
Riverside forest	10 to 20
Atlantic Landes	1 to 3
Intensive prairies	14 to 26
Humid prairies; sedges	around 1
Dry prairies, low steppes	1 to 3
Ruderal zones	4 to 30
Dunes	1 to 3
Halophytic formations	0.2 to 1
Cold regions	
Boreal and mountain forests	1 to 5
Elm forests	15
Alpine Landes	0.1 to 1
Alpine prairies	2 to 10
Tundra	0.03 to 0.5
Humid Landes and peat bogs	0 to 0.5

losses of water and thus great reductions of water potential. Often, the plant tolerates a temporary wilting and recovers very quickly as soon as it rains. The water use efficiency of plants varies widely depending on the species and especially on the metabolic type, more or less determined by the environment (Table 3.7).

There were early attempts to study the hydric relationships of plants (Tables 3.2 and 3.5) according to their biological type (in the sense of Raunkaier, or more recently of Lambinon) and according to the biomes studied in all the regions of the world. All the morphological and histological adaptations were observed, and cellular physiology was studied. It soon became apparent that **the variety of structural and physiological types is greater than the variety of habitats** and thus, in a given environment in a given place, **various strategies of adaptation** may be present. The notion of genotype emerges clearly. O. Stocker recognized that a critical evaluation must be made of the significance of certain characteristics considered ecologically important for life in a particular environment.

Among all these strategies of behaviour, different ecophysiological types and living strategies can be distinguished (Stocker, 1931; Korner, 1995). For a view of the problem as a whole, we can try to classify them according to the major biological types.

3.2. Ligneous organisms

3.2.1. Large trees

The organization of large trees and their sensitivity to embolism keep them from showing palpable variation in relation to an optimal hydric state. Any variation must be struggled against as soon as it is detected. Stomata respond to a low loss of water and temporarily reduce transpiration particularly during the hot hours. In spruce, for example (Fig. 3.18), a sort of distribution of spots is observed between the different parts of the foliage. The leaves in the shade close their stomata as soon it becomes hot. They thus integrally preserve the water reserves. The middle leaves that are exposed to sun have a classic behaviour with a reduction from the late morning, a compromise situation. The leaves at the top keep up a good rate of transpiration throughout the middle of the day. These leaves receive the most light and they must achieve the maximum rate of photosynthesis, which must be done very likely by exhausting the water from other parts of the plant. This is not common to all trees. The locust tree and others may allow their water potential to decline considerably. In the case of the locust tree, this is surely a key to its growth in difficult environments such as the slag heaps of Saint-Etienne.

3.2.2. Sclerophyte evergreen trees and shrubs

Most small trees and shrubs that are sclerophytes live in a Mediterranean climate, with a dry summer, or in an arid subtropical climate (African

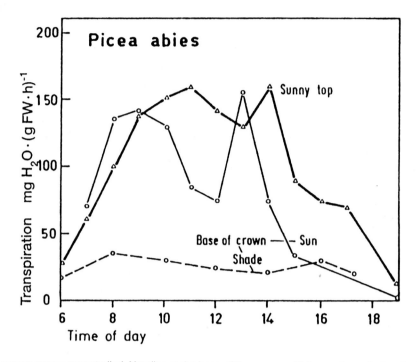

A young spruce was studied. Needles at the base of the crown, which are always in the shade, close their stomata early in the morning. Needles that are low but receive sunlight for some hours react irregularly. Needles at the top, which receive light all day, ensure an active and regular transpiration and thus photosynthesis.

Fig. 3.18. Daily fluctuations in transpiration from different parts of the foliage of a tree
(Pisek and Tranquillini, 1951, in Larcher, 1995).

Sahel, northeastern Brazil) with a short rainy season in summer and long, irregular dry periods. The plant formations are characteristic and are called Maquis or Chaparral.

There is a general adaptation to effect low but continuous photosynthetic activity (thus transpiration) with the help of a deep root system, for example (Fig. 3.19), except in extreme situations. Leaves that are greatly exposed to sun close their stomata very quickly but those in the shade close their stomata slowly, which allows low photosynthetic production. In terms of water content, stenohydric species can be found, such as *Arbutus unedo*, and euryhydric species such as *Olea oleaster*, or wild olive.

The water content (RWC), in proportion to the optimal water content, may decline in summer to 75% in the olive tree, while it remains at around 95% in *Arbutus*. In the olive tree, the leaf water potential may drop until the beginning of plasmolysis. There are three possible explanations for such behaviour. First of all, sclerophyte shrubs and trees have wood

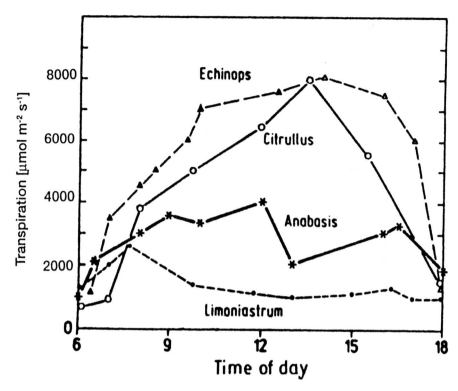

Echinops spinosus is a low sclerophyte shrub with deep roots. *Citrullus colocythis* is a geophyte with soft leaves but deep roots, both exhausting the water for quite a distance. *Anabasis aretoides* is a compact tuft plant, and its morphology limits transpiration. *Limoniastrum feei* forms a succulent rosette, and it can live in extremely dry places. It is considered to have very high cuticular resistance.

Fig. 3.19. Transpiration of Algerian desert plants, at the "beginning of the dry season" *(Stocker, 1974b, in Larcher, 1995).*

with very narrow and numerous vessels, which allows good conductivity of water as well as protection against embolism. The stenohydric plants have a conduction system that is never limiting, if there is water in the soil, and it is far away. The euryhydric trees, such as olive (Larcher, 1963), may show a high cellular resistance to the effects of dehydration during a rather wet year (50% of the damage intervenes for water deficits: 100 − RWC = WSD = 78%). In a dry year, however, the same effect is reached at around WSD = 85%.

3.2.3. Chamaephytes and tufted or prostrate plants in cold or alpine regions

These small plants are always in an open environment, greatly exposed to wind. The evapotranspiration is thus high even if the climate is not very hot. The plants have many short ramifications and a rather compact shape that is genetically determined. Such forms avoid breakage caused by snow and also allow conservation of internal heat, mostly in the tufted plants (*Androsace*) or prostrate plants (*Loiseleuria procumbens*).

Even if the stomata are highly sensitive to dehydration, the microclimate created by these plants often gives the leaves an environment with moist air and they can keep their stomata at least partly open without risk. Tufted plants have a very good capacity to absorb soil water. Their roots are thick and deep. In addition, because of their compact shape, with a sort of sponge at the interior created by the debris of old, dead leaves and all the dust captured between the small, rough branches, the tufted plants have a good reserve of moisture. Also, their shape ensures that the transpiration area is relatively small.

3.2.4. Extreme case of some semi-desert prostrate plants
A very few plants (Farrant et al., 2000) called "resurrection plants" living in South Africa are extremely euryhydric, by several means, such as the cell walls folding during severe dryness.

3.3. Herbaceous plants

3.3.1. Hemicryptophytes with large leaves
Hemicryptophytes with large leaves normally transpire a great deal because of a good root system and conduction. Those growing in a sunny and rather arid environment in particular can tolerate considerable dehydration, and the high transpiration (Table 3.5) may allow them to escape some of the sun's heat radiation and thus avoid possible thermal shocks (Muller-Stoll, 1935). This is the case of *Prunella grandiflora*, but others such as *Geranium sanguineum*, which has a superficial root system, very quickly close their stomata as soon as there is a shortage of water and wait for the dry period to pass.

3.3.2. Plants of the Megaphorbiae
Plants of the Megaphorbiae lose a great deal of water from their large leaves and often spectacular inflorescences. The soil water is never limiting, and the water content of the plants is often greater than 90%. However, on very hot summer days water deficits of the order of 20% may appear for a few hours, which is certainly the result of a limited flow of water in the conduction system (Morovoz and Belaya, 1988).

3.3.3. Geophytes
Geophytes are generally stenohydric or hydrostable. They usually develop in spring during a period in which the soil is very moist and not excessively hot. In summer they simply avoid water loss by allowing their leaves to dry up and fall and entering a period of dormancy.

3.3.4. Sciaphytes

Living in an often moist atmosphere, sciaphytes open their stomata very quickly during periods of dappled sunlight and close them equally quickly when they are in shade. Their cuticle, however, is very thin, and if they suddenly find themselves in sunlight for one or two hours, for example, after overhanging branches have broken off (see Chapter 5), their conducting system is reduced, and they rapidly dry out.

3.3.5. Grasses and reeds (hemicryptophytes and narrow-leaved therophytes)

Plants of "graminoid" bearing have overrun all lands and habitats from the water to the deserts. Certain species consume large quantities of water, particularly those found in marshes or on riverbanks. Their strategies differ greatly, from the stenohydric species to the euryhydric species presenting metabolic adaptations (C4 plants) or morphological adaptations, such as rolling of leaves.

3.3.6. Semi-parasitic plants

In principle, semi-parasitic plants meet their needs for organic matter by their own photosynthesis. If they are epiphytes, they draw from their host plant all the water they require, as with Loranthaceae, many species of which live on the branches of host trees. Other species live in the soil and, even though they have roots, probably draw a great deal of water directly from the roots of their host plants, as well as amino acids (e.g., *Rhinanthus*), because they seem deficient in enzymes such as GS and GOGAT.

All these plants draw their minerals from the host plant. It is for this reason that they pump an enormous quantity of water and transpire a great deal, the minerals exhausted from the soil already having been used to a great extent by the host plant. In consequence, the stomata of the members of the family Loranthaceae remain open for a very long time, even if the host plant has closed its own stomata following a hydric stress. The stomata of certain Scrophulariaceae remain open even when the leaves of the host plant have begun to wither (Shah et al., 1987).

3.3.7. Epiphytes

Epiphytes draw all they need from the atmosphere—water, minerals, and of course CO_2—by means of water vapour, aerosols, fogs, and dust. They are poikilohydric plants that can resist desiccation. Among them, the vascular plants have developed forms and tissues that allow them to capture water and store it in large quantities: e.g., the nest shape of some ferns, the dense base of bromeliads, and the velamen of orchids. In the misty mountain forests in the tropics (the *nebelwald*), the epiphytes, which represent 5 to 10% of the biomass, intercept around half of the water. Their highly exposed situation brings about a great deal of evaporation, particularly in the lichens.

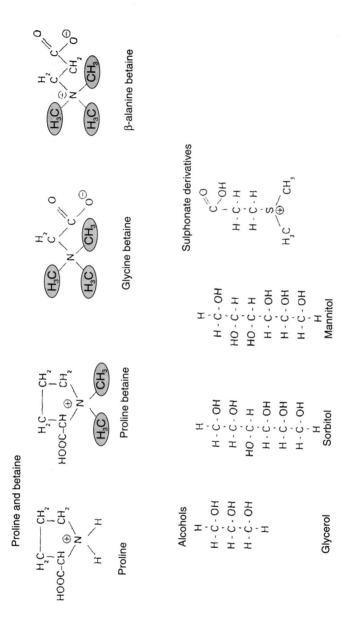

Fig. 3.20. Compatible substances in the cytoplasm during accumulation of NaCl in the halophytes
(Luttge, 1992).

These substances maintain the equilibrium of water potential of the cytoplasm with that of the vacuole and the apoplast. They also permit the enzymatic activities necessary to cells.

There are also physiological adaptations: for example, some bromeliads have crassulacean acid metabolism, which allows the closing of stomata during the day, and a great drought tolerance.

3.3.8. Halophytes

Even though halophytes are in general found in aquatic or semi-aquatic environments, their presence in saline water or soil has given them a peculiar physiology. Unlike many aquatic species, these plants contain a large amount of water and are relatively stenohydric. The water potential and the soluble salt content are close to those of the surrounding environment, more or less saline.

A halophyte of a relatively less saline environment, i.e., living in a soil fed partly by fresh water, such as *Cochlearia anglica*, is observed to have a high water content and a rather low salt content in the leaves, about 93 to 95% water and 5 to 15 g/l of Na^+ plus Cl^- per litre of water in summer (Leclerc, 1964). In *Obione* or *Halimione portulacoides*, which live in a more saline environment (mud invaded by salty marshes), the leaves in summer have 85 to 90% water and 25 to 30 g/l of Na^+ plus Cl^-. The salts accumulate in the cell vacuoles. The osmotic equilibrium in the hyaloplasmic compartment is ensured by the accumulation of oses, polyols, amino acids (most of all proline), or betains (Luttge, 1994; Fig. 3.20). These various organic substances are compatible with a good activity of enzymatic systems, even at potentials of –6 to –9 MPa.

Chapter 4

Ecophysiology of Development

1. INTRODUCTION

The fundamental characteristic of plants is their indefinite growth. They do not, in principle, stop growing at a given age, as do most animals. This indefinite growth is due to the presence of *subapical meristems*, cells of an embryonic nature that maintain the capacity to divide and produce new cells throughout the life of the plant. These cells *differentiate* to grow continuously *all the tissues* of the plant.

Subapical meristems, as their name indicates, are located very close to the extremities of the plant. The overall result of their activity is elongation of the plant by the formation of daughter cells. Lateral growth is clearly more limited, except in the formation of roots and lateral branches. The resulting organisms may be large and are often characterized by an elongated form with more or less lateral ramifications. After elongation, other meristems appear, the secondary meristems or layers (cambium, phellogen). These meristems result in greater lateral growth.

The indefinite growth of plants is subjected to regulation of endogenous origin and is also modified by exogenous factors. The interaction of endogenous and exogenous factors is a fundamental aspect of the ecophysiology of plant development. The overall regulation is based on patterns of organization identical to differentiation, leading to a wide range of ages in plants. Among those that can live for a long time are flowering plants such as *Androsace helvetica*, which never grows beyond a few centimetres, as well as some Australian *Eucalyptus* species that may grow to over 100 m.

Exogenous factors, of course, exert a fundamental constraint on growth and development, but it is not simply a mechanical constraint. **The plant perceives and reacts to its environment**. It receives signals from the outside world that it will modify, amplify, and interpret, and its reactions are well-determined and regulated on the scale of the cell and then the whole plant. There is thus an ecophysiology of stimulus and reaction.

On the basis of a certain number of internal correlations, the plant responds as a whole, as a function of seasonal changes or progressive modifications of its habitat. From the simple chain of primary signal, amplifications, secondary signal, and simple specific reaction, a complex perception is realized at several levels of the plant, as well as a complex reaction. This perception and reaction still need to be extensively studied.

2. EXTERNAL FACTORS AND HORMONES IN REGULATION OF DEVELOPMENT

It is possible in a preliminary approach to show some pairs of external stimulus and action of a plant hormone (Fig. 4.1). Every part of a plant is well informed of the existence of an external stimulus by the presence of a plant hormone, which supposes synthesis and transport of this plant hormone, possibly in synergy or in antagonism with another plant hormone.

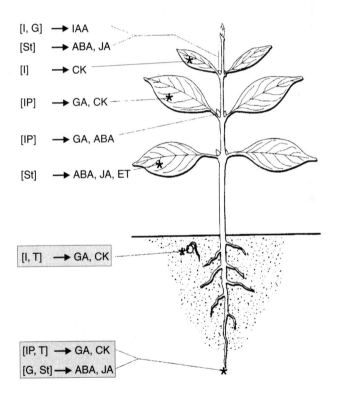

I, light (quality, intensity). T, temperature. ABA, abscisic acid. GA, gibberellic acid. IP, photoperiod. St, abiotic stress. JA, jasmonic acid. ET, ethylene. G, gravity. IAA, auxin. CK, cytokinins.

Fig. 4.1. Phytohormones and environmental stimuli *(Larcher, 1995).*

The pairs observed vary according to the species or groups of species, stages of development, target organs, seasons, and other factors. The result is an integration of the synchrony between the development of the plant in the course of a year and the entire surrounding microclimate.

Other integrations between groups of reactions and groups of stimuli may be observed in orientation in space, in development of aerial and underground parts, or at the trophic level in relationships between source organs and sink organs.

Among the external effects, the effects of light are particularly important (Table 4.1).

Table 4.1. Effects of light on plant development
(Salisbury, 1985; Krunenberg and Kendrick, 1986)

Phenomenon	Mode of action	Spectrum	Type of period	Delay in response
Seed germination and emergence of buds	I	R/FR, B	P	hour or day
Stem elongation	Q, F	R/FR	P	min.
Orientation of stems	Q, F	B		min.
Orientation of leaves	Q	R/FR	C	min.
Blooming of flowers	I	R/FR	C	hours to weeks
Development and filling of sink organs	I	R/FR	P	
Synthesis of enzymes	I	R/FR		hour
Activation of enzymes	I	R/FR		min.
Dormancy	I	R/FR	P	
Membrane potentials	I	R/FR		sec.

I, inductive. Q, quantitative. F, formative. B, blue light. R/FR ratio of red to far-red. P, photoperiod. C, circadian rhythm. The delay in response is the interval between the beginning of illumination and the beginning of reaction.

2.1. Role of light

2.1.1. Introduction

Light is a complex stimulus, in itself by its range of wavelengths and in the variety of its effects beyond photosynthesis: light can stimulate movement, synthesis, and complex phenomena of development and differentiation from the molecular scale to the entire organism (Table 4.1).

Light acts by the intermediary of antennae that receive the light signal. These antennae are photoreceptor pigments. At another level, light acts also in interaction with endogenous rhythms: this results in photomorphogenesis and photoperiodism.

The effects of ultraviolet light (Fig. 4.2), blue light, and especially red light, by the intermediary of photoreceptor pigments, are known. Such pigments are complexes with FAD (flavin adenine dinucleotide) or FMN (flavin mononucleotide), phytochrome, or cryptochrome. These photoreceptors are inserted all over the membrane and allow the plant to

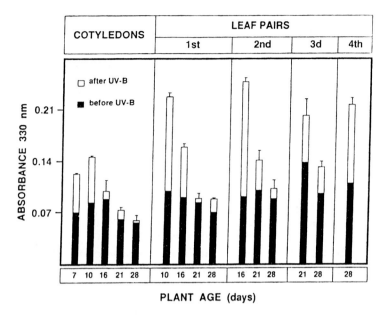

The columns represent the absorption of ethanolic extracts, at 330 nm from the cotyledons (day 7) to different pairs of leaves (no. 1 to no. 4) that are well developed and spread out, in the order of their appearance (from day 10 to day 28), on different ages of plants. The black part of the columns represents the non-irradiated plants, the white part represents plants after 24 h irradiation. The data are from three plants, thus from six leaves, and the standard deviations are indicated.

Fig. 4.2. Accumulation of flavonoids, essential or induced by UV, as a function of the development of *Arabidopsis (Lois, 1994).*

recognize the direction of light. This recognition stimulates movements, so that the plant better exposes itself to light or avoids light.

Very slow movements, detectable after a few hours, result from oriented growth called *phototropism*: positive phototropism for the plant to face and grow towards the light in the case of above-ground parts of higher plants, or negative phototropism in the case of mushrooms.

Faster movements, detectable within a minute to an hour, result from photonasty (made possible by variations in petiole turgor) or even from "solar tracking" (following the direction of the sun). A well-known example is the orientation of leaves of *Dryas octopetala* observed by Svoboda (1977) in the Baffin Islands (north of Canada), by which the plant captures the heat of the sun and optimizes its photosynthesis.

2.1.2. Photomorphogenesis

Photomorphogenesis can be defined in a broad sense. It is observed on the molecular and cellular scale. Light acts by the intermediary of a photoreceptor that gains the energy of a photon and uses it to change conformation of itself or another molecule, which induces the formation of secondary messengers: cyclic AMP, Ca^{+++}, and its regulation by calmodulin.

Genome activity is therefore regulated, and many enzyme activities, for example stimulation of Rubisco activity, are directly or indirectly regulated. Also, photoperiod control of cell differentiation and tissue development is observed, which allows photoinduction of germination in higher plants. The photoreceptor is also called phytochrome. This phytochrome is active either in the form P_r, which absorbs red light (most active λ: 660 nm) or in the form P_{fr}, which absorbs far-red light (most active λ: 730 nm).

The equilibrium of the two forms thus depends on the quality of the light, more or less rich in red or far-red light. The requisite P_r/P_{fr} equilibrium depends on the species and the phenomenon to be activated. There are species of well-illuminated environments that germinate only in light and when the light has a ratio of red to far-red (R/FR) of 2 to 3.

The effects of phytochrome interfere with other environmental influences such as vernalization (Fig. 4.3).

Species of a dark environment germinate at R/FR values from 0.1 to 0.5, values found in the forest undergrowth where the canopy has intercepted a major part of the red radiation. In these species, photodormancy is generally observed when R/FR is 2 to 3. In some cases, seeds may form in an environment rich in far-red light, but they lie dormant and buried, and they germinate only much later, when they ultimately find themselves exposed to red light. This is the case, for example, with digitalis (*Digitalis purpurea*) following an opening up of the canopy in a forest after a tree falls. Photodormancy helps regulate the generation of subsequent plants.

Sometimes, photoinduction of germination is modulated by the temperature. In some species presenting photodormancy, a low temperature may increase sensitivity to light, and the plant thus requires less light intensity to reach dormancy.

In the context of interactions between light and endogenous rhythms, photoperiodism is observed and is markedly seasonal in upper and middle latitudes.

2.1.3. Seasonal photoperiodism

The photoperiodic signal is fundamentally linked to the science of astronomy. It allows detection of the relative length of days and nights, and thus short, medium or long days. It is in principle independent of the temperature and more generally of the present climate. The photoperiodic signal is coordinated with endogenous circadian or ultradian rhythms. It may also, in conjunction with earlier signals, alert the plant to future environmental stresses.

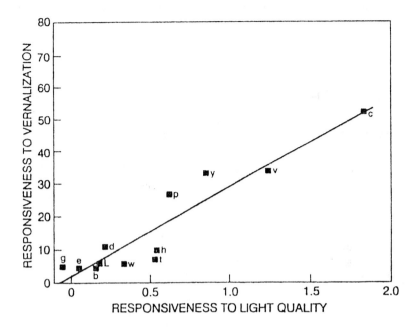

Several mutants of *Arabidopsis thaliana* (L.) Heynh. race Landsbery *erecta* were tested. It was previously shown that the number of leaves of mutants insensitive to R/FR does not change when R/FR changes from 1 to 2.8. On the other hand, in sensitive mutants, the number of leaves increases with R/FR. The X-axis represents the relative increase in number of leaves with a change from R/FR = 1 to R/FR = 2.8. The mutants most sensitive to vernalization (e.g., c and v) are also the most sensitive to the R/FR ratio. The coefficient of correlation r^2 is 0.88. The response to vernalization is measured by comparing the number of leaves at the beginning of flowering in non-vernalized plants (which have more leaves) and vernalized plants (20 days in dark at 2 to 4°C). The percentage of decrease in number of leaves in vernalized plants is represented on the X-axis.

Fig. 4.3. Response of late-flowering mutants of *Arabidopsis* to quality of light and to vernalization *(Bagnall, 1992)*.

Seasonal photoperiodism primarily affects germination. It also affects seedling emergence, leaf development, ramification, development of reserve organs, and especially induction of flowering (Table 4.2). An example of an alert on future stress is found in some plants of a Mediterranean climate in which seasonal dimorphism is observed, essentially in the leaves. The appearance of long days indicates to the plant the beginning of summer, and thus the dry season. The plant reacts either by forming new leaves adapted to aridity or by shedding leaves.

In the higher plants as well in cryptogams, photoperiodic signals allow the **precise seasonal programming** of reproduction—flowering, pollination, hardening of fruits, and fruit dispersal—in an optimal fashion according to the average climate observed by the plant and by preceding

Table 4.2. Responses of plants to photoperiodism (Salisbury, 1982, in Larcher, 1995)

Development activity	Example of plant species	Favoured by short or long days
Induction of seed germination	*Betula pubescens*	LD
	Nemophila menziesii	SD
Induction during seed development	*Chenopodium album*	SD
Stem elongation	Almost all plants	LD
	Alstroemeria cv *Regina*	SD
Suckering in cereal grasses	*Hordeum vulgare*	SD
	Oryza sativa	LD
Ramification	*Oenanthera biennis*	SD
Leaf enlargement	*Glycine max*	LD
Leaf succulence	*Kalanchoe blossfeldiana*	SD
Development of reserve organs	*Solanum tuberosum*	SD
	Allium cepa	LD
Flowering	*Rudbeckia hirta*	LD
	Cosmos sulphurus	SD
Promotion of female flowers	*Cannabis sativa*	SD
	Spinacia oleracea	LD
Clonal reproduction	*Bryophyllum*	LD
Leaf abscission	Deciduous trees in high and medium latitudes	SD
Bud emergence	Deciduous trees in high and medium latitudes	LD

generations living in the same place. Thus, we observe ecotypes, or species, linked to particular photoperiods. Photoperiodism is thus a source of biodiversity and, in the framework of reproductive strategy, it could lead to **coevolution** between, for example, plant and pollinator.

2.2. The role of temperature

2.2.1. Introduction

Temperature, of course, has a general effect on plant metabolism. Conventionally, biochemical reactions have an energy of activation and, in consequence, rates of reactions increase with the temperature, according to a Q10 (relative increase in activity if the temperature increases from 10°C, which is often of the order of 2 to 3), which derives from the Van t'Hoff Law.

Unlike what is observed in chemistry, the reactions of living things increase only up to a certain temperature, which is considered optimal. Subsequently, the rates of reactions rapidly diminish, essentially because of denaturation of enzymes and destabilization of the membranes.

Also, many threshold phenomena are observed in biology, partly linked to the membrane physiology. Sometimes, there is a clear thermoinduction, in which the attaining of a threshold will abruptly trigger an entire series of actions. Temperature thresholds and thermoinduction depend greatly on plant groups and especially on living conditions of different species.

It is also possible to detect manifestations of thermoperiodism. A daily or seasonal alternation of relative heat and cold, in most cases, is useful to the plant.

2.2.2. Temperature and germination

Some effects of temperature can be described with respect to two aspects of plant development: germination and formation of reproductive organs. Generally, germination requires that the temperature go beyond a certain threshold value or sometimes drop below a threshold. Once the threshold is crossed, the rate of germination grows exponentially with the temperature up to a maximum, then there is a decrease due to a set of lethal and sublethal phenomena and significant catabolisms that intervene at extreme temperatures, which have been observed experimentally but occur less often in natural conditions.

Table 4.3. Minimum, optimum, and maximum temperatures (°C) of seed and spore germination (data from many authors, in Larcher, 1995)

Taxonomic group	Minimum	Optimum	Maximum
Fungi			
plant pathogens	0–5	15–30	30–40
soil fungi	~ 5	~ 25	~ 35
thermophilous fungi	~ 25	45–55	~ 60
Monocotyledons			
prairie grasses	3–4	~ 20	~ 30
temperate cereals	sometimes 0, 2– 5	20–25, sometimes 30	30–37
rice	10–12	30–37	40–42
tropical and subtropical C4	sometimes 8, 10-20	32–40	45–50, sometimes 40
Herbaceous dicotyledons			
plants of high altitudes and tundra	sometimes 3, 5–10	to 20	
prairie plants	sometimes 1, 2–5	15–20	35–45
temperate cultivated plants	1–3, sometimes 6	15–25, sometimes 30	30–40
subtropical and tropical cultivated plants	10–20	30–40	45–50
Desert plants			
summer germination	10	20–30, sometimes 35	
winter germination	0	10–20	to 30
cactus	10–20	20–30	30–40
Trees (after stratification)			
conifers	4–10	15–25	35–40
angiosperms	< 10	20–30	

The ranges are often wide because several factors have an impact.

In this context, there are differences among species, depending on their geographic distribution (Table 4.3) as well as on particular ecosystems. Ecotypes have particular modalities of germination as a function of the temperature. The same is true of cultivars (Table 4.4). In the framework of microevolution, human activities have triggered the sometimes rapid appearance of variants having a particular physiology at the stage of germination.

For example, I. Till showed that several ecotypes of *Poa annua* were found to be selected on golf courses in California (Wu et al., 1998). In the parts that were well-watered even in summer, this grass germinated well in the summer. On the other hand, on the parts of the course that remained dry during the summer, a variant was found in which germination was inhibited by summer temperatures. It could germinate well (in terms of activity and rate) only in low temperatures.

Table 4.4. Germination as a function of temperature and varietal comparison
(Blumenthal et al., 1996)

Genotype	Temperature (°C)							
	5	10	15	20	25	30	35	40
A								
Grasslands Goldie	93.8	87.5	96.8	97.2	94.7	84.7	84.9	74.5
Grasslands Maku	33.6	96.0	97.6	100.0	99.3	100.0	98.3	70.8
G4703	5.0	81.7	81.3	82.1	85.7	86.5	84.9	67.1
G4704	9.4	96.6	100.0	98.5	99.3	98.1	95.2	77.5
K × 1000								
Grasslands Goldie	6	7	25	31	28	9	7	16
Grasslands Maku	1	3	18	26	33	46	36	10
G4703	2	2	11	14	15	16	16	10
G4704	2	3	13	20	23	15	12	6
T_0								
Grasslands Goldie	53.4	50.9	17.9	16.5	13.0	1.1	22.3	26.7
Grasslands Maku	63.6	74.2	24.8	19.8	16.7	15.9	16.2	29.7
G4703	93.3	82.7	25.4	17.3	14.8	14.0	17.1	34.0
G4704	57.0	85.9	31.3	23.1	20.3	18.5	22.8	33.2
T_{50}								
Grasslands Goldie	163.4	215.2	46.0	39.4	39.2	79.9	128.2	70.4
Grasslands Maku	641.1	285.9	62.8	46.5	37.5	30.9	35.6	99.4
G4703	555.3	544.7	86.6	67.9	63.9	58.5	61.8	105.8
G4704	365.0	374.6	86.0	57.2	50.1	66.4	79.3	207.3

A, germination activity, counted on 34 days and sometimes reaching 100%. K, germination rate constant. T_0, time (days) at the end of which the first germination is observed. T_{50}, time at the end of which 50% of seeds that can germinate have germinated. The percentage of germination at time t follows a Mitscherlich function of the type $p = A \{1 - \exp[-k(t - t_0)]\}$. Grasslands Goldie and Grasslands Maku are two genotypes of *Lotus corniculatus* (birdfoot trefoil) of New Zealand. G4703 and G4704 are two genotypes of *Lotus uliginosus* (greater birdfoot trefoil) of New Zealand.

A related phenomenon is temperature-dependent **dormancy**. For a long time, the technique of **cold stratification** has been practised, by which the dormancy of winter cereals is lifted by low temperatures applied for several weeks (about 5°C for 5 weeks). On the contrary, treatment of certain varieties of rice seeds at temperatures close to 40°C rapidly lifts their dormancy.

During germination a heterogeneity of behaviour appears between the different parts of the young plant. Whereas cell division as well as cell enlargement demand a good deal of heat, cell differentiation can easily occur at lower temperatures.

In a temperate environment, root development (elongation, differentiation) can be very active from 2 to 5°C, which is helpful if the spring is feared to be relatively dry. The possibility of root growth at low temperatures is useful in horticulture for the purpose of transplanting and budding. Finally, as a general rule, in the course of vegetative development of the young plant, a **circadian thermoperiodism** is observed. It follows overall circadian variations of the temperature of biotopes. For example, in some species of cactus, a difference of 20°C between day and night is required for optimum growth. Many plants of a temperate environment require a difference of 5 to 10°C between day and night.

2.2.3. *Temperature and reproduction*

Vegetative or sexual reproduction is also closely dependent on temperature. In a cold climate, there is a tendency toward vegetative reproduction. There is also a seasonal effect. Certain mountain plants reproduce sexually in summer, when insects are active and when pollen germinates well, while they propagate vegetatively in the early autumn. Behind all these reproductive phenomena, there are links between temperature and hormonal activity.

Purely in terms of sexual reproduction, a very important thermal factor has an impact from the outset: flower formation, in most cases, is induced only within a narrow margin of temperatures, depending on species and varieties. This is **vernalization**: it requires several weeks of temperatures of around 5°C (between –3°C and 12°C) for flowers of cereals, biennial plants, and some fruit trees such as peach to be initiated at the subapical meristems.

We may also find sequential programmes of flower formation and development, particularly in geophytes with bulbs in continental temperate climates. This is the case, for example, with tulip and hyacinth. In hyacinth, flowers are initiated by temperatures of over 30°C for a few weeks; subsequently (in a natural environment), differentiation of tissues and flower organs occurs if there are temperatures of 10 to 15°C for around two months. Finally, a temperature of about 15 to 25°C is required for elongation of flower tissues, and then proper flowering occurs.

Once the flower is operational, pollen germination is temperature-dependent. For example, the difficult germination of pollen of lime tree (*Tilia cordata*) at low temperature limits the distribution of this species towards the north.

Temperature and light have an impact on the different life stages of the plant, from embryo to senescence, in relation with hormonal activity.

2.3. Hormones and growth regulation

2.3.1. Introduction

A genetic programme of development lays down the basic scheme that will be followed throughout the life of a plant. This programme is modulated by external conditions, in relation with hormones that ensure, in particular, coordination between the various organs.

The life span thus programmed is effectively quite long (Table 4.5). The reproductive stage, considered adult, may be reached slowly, and the sexual life span may be very long. For example, a *Pinus aristata* (bristlecone pine) of the White Mountains of California still produces cones at an age of more than 4000 years.

Each stage of development has specific environmental requirements and hormonal activities. The previous stage always has a marked influence on the current stage. The previous generation also has a significant effect on the development of male and female gametophytes (pollen grains, embryo sacs), ovules or ovule equivalents, and embryos and seeds.

2.3.2. Embryonic phase

The mother plant transfers nutrients to the gametophytes and then to the embryos and seeds, and it transfers hormones, including indoleacetic acid (IAA), gibberellic acid (GA), cytokinins (CK), and abscisic acid (ABA).

Depending on the external conditions of life of the mother plant, the ovules and embryo sacs are more or less numerous and more or less developed. The same applies to pollen, in rice, for example (Tani, 1978).

When the embryo differentiates and develops, the seed is formed and filled with food and energy reserves. Sometimes the seed matures and disperses before the embryo is developed. This is often the case in Ranunculaceae (e.g., ficaria), and also in some Apiaceae (berce: *Heracleum*) and some trees such as ash.

During the creation of the seed, the mother plant has an important effect on the teguments, which may serve to filter light (R/FR ratio). Teguments can also be charged with inhibitors that are more or less water-soluble, or even sclerotize to a great extent.

If the mother plant is in good health it produces large seeds, which will ensure better germination and emergence.

Table 4.5. Life span and sexual maturity in years
(data from many authors, in Larcher, 1995)

Plant	First flowering (years)	Life span (years)
Annual plants	some weeks	some months
Biennial plants	1	often 2
Herbaceous perennials	2–10	10–40
(mostly hemicryptophytes)		
Chamaephytes	5–10	50 or more
Shrubs (1 to 3 m high)	5–20	50–100 or more
Palms	up to 50–80	50–100
Dicotyledon trees, pioneers and/or nomads or growing on riverbanks		
Wild cherry	to 5	60–70
Alder	to 10	80–150
Poplar, willow	5–15	80–150
Birch	5–20	100–150
Ash	10–40	100–250
Apple, pear	5–15	100–250
Locust tree	10–20	100–300 approx.
Trees mostly at end of plant succession		
Elm	15–30	200–400
Maple	15–30	150–500
Beech	30–80	300–900
Lime tree	15–25	700–2000
Oak	20–50, sometimes 75	500–2000
Conifers		
Pines of subtropical regions	5–8	100–300
Pines of temperate regions	10–20 (40)	300–500
Spruce	30–50	200–500
Larch	10–20	200–800
Douglas fir	15–20	500–1500
Yew	10 and above	up to 2000
Juniper	10–20	300–2000
Cypress and other Cupressaceae	10–20	300–2000
Sequoiadendron giganteum	15–20	2000–4000
Pinus aristata (Rocky Mountains)	?	2000–4000
P. aristata/longeavia (White Mountains)	?	up to 4800

The data represent wide ranges due to the number of species in a genus. It must also be noted that in many trees the age of flowering is greater if the population is dense.

The embryo also manufactures its own hormones, which modulate the metabolism of the mother plant and bring about the adjustments needed to regulate the number of fruits and their maturation. ABA has an effect by provoking either early abscission of very young fruits (if there are too many fruits for the prevailing conditions) or abscission at the end of the normal term of development.

2.3.3. Germination
Germination is considered to last from the imbibition of the seed to the emergence of the radicle. Even if there is imbibition (and thus good

hydration), germination cannot be triggered immediately because in many cases dormancy must have ended. Dormancy ends:

- when an internal biological clock indicates a favourable phase (with cyclic return),
- by a particular exposure to light,
- by stratification,
- following heavy rain,
- after the prolonged action of bacteria and/or fungi on the teguments, and
- after the passage of fruits through the digestive tract of an animal, which is one of the bases of **zoochory**, which enables the dispersal of a species.

Once dormancy has ended, and if there is good hydration (imbibition), germination begins, induced by the activity of enzymes already present or the activity of the genome: the cycle of pentose phosphates gets under way, as well as glycolysis and various hydrolase activities (e.g., amylase, lipase). For example, GA is observed to induce the synthesis of enzymes in barley, in cells with grains of aleurone. CK and IAA favour cell division and enlargement.

The rapidity of germination depends on the conditions. It will be high, for example, if the risks of drought are high. The radicle must in that case form and emerge quickly in order to search for water deep in the soil. It could be quite slow in species in which the seeds must complete the maturation of the embryos. Once the seedling is established with its roots in the soil, emergence begins.

2.3.4. Emergence
Emergence corresponds to the coming out of the aerial organs and is accompanied by root development. The equilibrium of root/stem development is essential. Emergence can be said to end when the young plant becomes completely autonomous with respect to energy, and thus independent of the mother plant in terms of food.

It is the most vulnerable phase in the development of the plant. The young aerial part is easily browsed by herbivores. The cells walls of the young organs are thin and can be attacked by fungi. Moreover, the soil surface may be a difficult medium: if the soil is dark and receives much sunshine, the surface may heat up to more than 50°C in certain conditions in a temperate climate. If the soil is saline, salt tends to concentrate at the surface. Most of all, the emerging plant is highly sensitive to drought, and the root/stem ratio is thus critical.

On the one hand, root development must be large enough to support the young plant, which requires an energy investment of part of the seed reserves. On the other hand, the aerial part must also be well developed (in the beginning from the seed reserves) to ensure sufficient photosynthesis to take over from the seed reserves in providing energy.

The takeover is not evident for all species at the end of the emergence period. In some cases, there is a crisis. In the young maize, for example, for one or two days photosynthesis ceases to increase (Fig. 4.4). The plant has two or three leaves, and there should be no problem, but it also forms a network of adventitious roots, which consume the last reserves of the seed and may also exhaust the mineral and organic nutrients of the leaves at the cost of photosynthesis (Leclerc and Abd el Rahman, 1988).

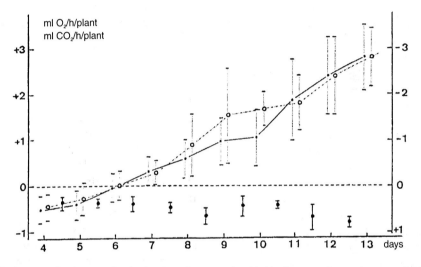

The increase in photosynthesis shows a plateau at the third-leaf stage, after exchanges of CO_2 (open circles) as well as after exchanges of O_2 (closed circles).

Fig. 4.4. Crisis of photosynthesis in very young maize
(Leclerc and Abd el Rahman, 1988).

For all these reasons, emergence is the great demographic problem. Its success is critical to the future of a given plant population.

3. DEVELOPMENT AFTER GERMINATION

3.1. Period of vegetative growth

In addition to the root/stem equilibrium, the possibility of rapid elongation of the aerial parts is often of critical importance, horizontally in the case of vegetative multiplication from stolons, and at the same time vertically to reach better light, in an environment in which there is competition from larger plants.

It is mostly in this period that **phenotypic plasticity** manifests itself in the adaptation of a plant to conditions of its habitat. The shape of plants, particularly trees, will often change characteristically with age, as with the Atlas cedar (*Cedrus atlantica*).

Independently of the trophic aspect, vegetative growth depends essentially on cell enlargement and differentiation. From the apical meristems, where cells multiply, cells begin to differentiate according to their position with respect to the principal axis (distance to the apex, angular position). However, there will be further cell divisions that give rise to various **primordia**: cell masses that will increase and differentiate to produce the various organs.

The order, arrangement, and nature of primordia are defined by a **plastochrone system** that determines the primary vascular vessels, angular distance between two successive leaves, and other characteristics.

Heat is an important factor that affects cell division, and light plays a major role in cell enlargement, which is manifested mainly by **elongation**.

The number of cells in a leaf seem to be determined overall by that of the primordium. If, however, conditions are unfavourable—intense browsing, drought, stunting, strong wind—the number of cell divisions and thus the number of cells will decrease. The leaf will be smaller essentially because it has fewer cells, not because the cells are smaller.

From the end of the divisions to the beginning of differentiation, cells enlarge some tens of times in volume. The elongation and general extension of cells depend on the interaction of receptors of blue light, phytochrome, turgor, acidity of the apoplast, and hormones. A threshold pressure Ψ_p^{lim} can be defined by which the cell wall is mechanically extended, irreversibly. In several cultivated plants it is around -0.5 MPa (Terry et al., 1983). For Ψ_p, the effective turgor pressure, to be greater than Ψ^{lim}, the plant must be well supplied with water and also photosynthesize efficiently, ensuring the uptake of new wall material and of organic solutes to reduce the water potential. The relative rate of growth (Lockhart, 1965) can be expressed as $dV/dt \cdot 1/V = m(\Psi_p - \Psi_p^{lim})(s^{-1})$, where m is a constant of plasticity of the wall, and V is the volume.

In the event of drought, cell division as well as cell enlargement may be affected by the absence or weakness of turgor, or low plasticity of the wall. Abscisic acid probably intervenes at several levels in these various aspects of reaction to drought.

3.2. Development during reproductive phase

3.2.1. Endogenous and exogenous factors

The reproductive phase is reached when the plant begins to flower, which happens after it has reached **flowering capacity**. Flowering capacity depends on a series of endogenous factors (fundamentally of the genome,

Fig. 4.3, and its expression), as well as the environment. It is made possible by **flowering induction**.

A plant can start flowering at a particular age (Table 4.5) or when it has formed a certain number of leaves and leaf primordia, or when it reaches a certain size. Although endogenous characters are involved, the environment has an influence in terms of nutrient supply. There are other factors that are **exogenous** in many species and have an **essential impact**, even if the plant seems to have reached a sufficient biomass in terms of nutrients.

In relation with phytohormones, light has an impact, as do temperature, possibly the hydric state, and sometimes a situation of environmental stress. Many species are observed to have a photoperiod (short-day or long-day plants) and/or a thermoperiod (e.g., hyacinth, which needs a hot period before flowering induction).

Even when development of an individual does not seem sufficient, flowering may occur as a response to a stress situation. This was observed, for example, by Yersin during culture assays of cinchona in Indochina (Bernard, 1955). The interaction of endogenous and exogenous factors manifests itself overall in that, although it determines the **quality**, that is, the possibility of making flowers and fruits, it also affects the **quantity** of flowers and fruits, which is chiefly influenced by the environment. The quantity is what is expressed (in terms of biomass) because it is the **reproductive effort**, which varies with the species, groups of species, and constant or random external factors.

3.2.2. *Reproductive effort*

Reproductive effort is measured by the percentage of total biomass produced by the plant, which is invested in the flowers and fruits by the effect of complex **source-sink relationships**. The active leaves (by the export of photosynthates) and senescent leaves (by cycling of proteins and destroyed polysaccharides, in the form of small molecules) are the source of organic nutrients, and the roots are the source of minerals. In ligneous species, the organic and mineral reserves of the branches and trunk also constitute sources. The flowers and then the fruits are the sinks.

The reproductive effort (Table 1.10) is highly variable, and it expresses in fact the result of competition between formation of the vegetative apparatus and formation of the reproductive organs. As a general rule, the reproductive effort is calculated for a single sexual reproduction. In many species, and groups of species, there is **competition** between sexual and vegetative reproduction.

Some entire groups have a low reproductive effort, for example, a large number of thallophytes. Lichens and certain mosses are able to colonize some of the most hostile environments in the living world by expending the minimum of energy for sexual reproduction.

On the other hand, the reproductive effort may be very large in many fruit trees that are **selected** for their fruits: some *Citrus* have an effort greater than 50%.

Average values are observed in the forest, for beech for example (10–20%) or for conifers (5–15%). In conifers, the reproductive effort comes from a good part of the seed reserves, even if by definition there is no fertilization. In an environment uncongenial to photosynthesis, for example the undergrowth (Table 1.10), the reproductive effort of many species is quite low. Reproductive effort is also linked to capacity for seed dispersal, often limited in the plants of the undergrowth (Fig. 4.5).

Reproductive effort is often seasonal (hot season in temperate environment, wet season in tropical environment). It also varies, in some cases, according to the year, the causes being either endogenous or exogenous. There are monocarpic plants, perennial plants that flower only

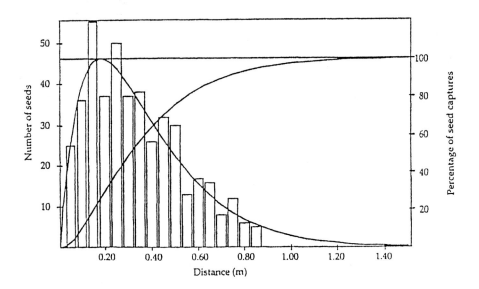

Erythronium grandiflora is a Liliaceae growing in the upper parts of the Rocky Mountain forests in North America. A vicariant species exists in France, *Erythronium dens-canis*, which also lives in the mountains, but at a lower altitude. *Erythronium grandiflora* produces seeds that are too heavy (0.16 ± 0.04 mg, n = 50) to be carried by even strong winds. The seeds have no appendages that allow them to be carried by wind. Nevertheless, wind is the only means of seed dispersal and carries the seeds for short distances (mean distance travelled, 0.325 m). The curve represents the frequency distribution of presences as a function of distance classes from the mother plant.

Fig. 4.5. Seed dispersal in *Erythronium*
(Weiblen and Thomson, 1995).

once in their lives. There are many examples, particularly in the taiga forests, of trees that flower and fruit only during certain "good years", which return cyclically. For example, in Siberia, *Abies sibirica*, *Pinus cembra*, and *Larix daourica* fruit once every 5 to 10 years, depending on the region. This has serious consequences for the demography of crosshills, for example, and causes the migration of large populations of squirrels.

3.2.3. Reproduction and biological types

3.2.3.1. Therophytes
In annual plants, the vegetative and reproductive development phases are often successive. For example, in *Arabidopsis thaliana*, a rosette of about 20 leaves is formed, then heading takes place, then formation of the floriferous stem. During the course of flowering, the senescence of the rosette begins (Blaise et al., 1998).

In *Arabidopsis*, as in many other species, it is the upper parts of the vegetative organs, close to the reproductive organs, as well as the peripheral parts of the reproductive organs (bracts, peduncles, glumes) that provide the nutrients essential to the seeds that will form. For example, it is the last leaf of wheat (the flag leaf) that nourishes the ear (Fig. 4.6).

3.2.3.2. Biennial plants (part of hemicryptophytes)
Biennial plants have a strategy similar to that of *Arabidopsis*, but over a longer period of time and with a wintering period after construction of the rosette. During the first year the rosette allows reserves to accumulate at the roots and in a thick but short stem. In the second year, the floriferous stem emerges and produces a few more leaves, as in *Digitalis purpurea*.

A second floriferous stem may emerge in the third or even the fourth year. There are multi-annual herbaceous plants with longer life spans, and a single individual may produce variable yields of flowers and fruits, depending closely on the climate of the preceding year, which may have been more or less favourable to the accumulation of reserves.

3.2.3.3. Hemicryptophytes and chamaephytes
In cold environments, many herbaceous species are multi-annual, and many species are ligneous while remaining moderate in size: these are **chamaephytes**. They can have a moderate reproductive effort because they are multi-annual. The plant makes a significant quantity of seeds only during favourable years so as not to put at risk the demography of the population.

Moreover, in a hostile environment, a tendency to vegetative multiplication is observed, by means of stolons that are sometimes very short, allowing, for example, formation of a new rosette for the same clone almost at the same place, if the soil is good. In the mountains, a plant is

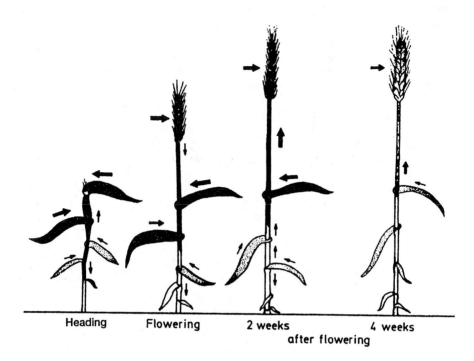

Heading   Flowering   2 weeks   4 weeks
                                  after flowering

It is always the youngest leaves (shown black in the figure) that produce more. At first the photosynthates nourish the growing terminal part and ultimately the ear is by far the principal sink. Some photosynthates go to the roots. Finally, the last leaf (flag leaf) nourishes the seeds as they are filling and maturing. The arrows indicate the direction and intensity of transport.

Fig. 4.6. Production and movement of photosynthates towards sinks during the development of wheat *(Stoy, 1965, in Larcher, 1995)*.

often observed to multiply sexually in full summer, when the days are long, but the same plant changes to vegetative multiplication (viviparity in Poaceae, for example) at the end of the summer when it is colder, the days become shorter, and fertilization by insects is more difficult.

In phanerophytes the same tendencies are seen as in the chamaephytes, with many fruits in good years, if reserves could be accumulated in the preceding year and, on the other hand, formation of predominantly or exclusively vegetative buds in a bad year or a succession of bad years.

The root/stem equilibrium changes greatly with age (Fig. 4.7).

4. SENESCENCE AND RECYCLING PHASE

Old age affects plants as well as animals. It is not always evident. Sometimes it is visible and sometimes not. Senescence is clearly seen in

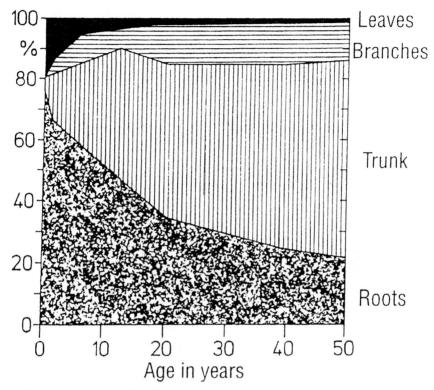

In the beginning, the allocation of biomass in the roots is considerable, corresponding to the need for the young tree to reach stable water resources deep in the soil, which will be useful in the future. With age, the proportion of biomass in the trunk becomes predominant and constitutes a sink for CO_2 passed into organic form, to be managed with caution.

Fig. 4.7. Accumulation of biomass in tree trunk
(Remezov and Bykova in Mitscherlich, 1970, reprinted by Larcher, 1995).

the therophytes. It is less easy to detect in bulbous geophytes, in which the bulb renews itself, or in many chamaephytes, which develop to a large extent by vegetative multiplication, and in which the notion of the individual is diffuse. In phanerophytes (trees), particularly conifers, the slowing of cambial activity, apical elongation, and fruit production are not easily distinguished from the effects of bad years.

To simplify, senescence can be said to begin when the apical meristems, independently of environmental conditions, slow their activity and begin to degenerate. It is a phase that may be very long in the trees. In certain species of great longevity (Table 4.5), the senescence phase probably exceeds half the lifetime.

Senescence must be considered on different scales, of the whole plant, organs, tissues, and cells. Cells (piliferous epidermis of roots, cells of the lining) or tissues (epidermis of young stems and roots, xylem), or even organs such as cotyledonary leaves live for a very short time. In deciduous trees, leaves live for only one season, as do fruits, very often, and senescence is considered to have ended with abscission.

Finally, we must consider senescence on the scale of settlements and communities. For example, pioneer settlements are young and primeval forests have a large proportion of old trees.

Senescence is programmed on various scales, from the cell to the organs, or even large parts of an individual, in relation with endogenous and exogenous factors.

At the cell level, two major situations are found. Senescence may be the final part of the programme of cell differentiation, as with cells of the xylem or sclerenchyma. Or it may act after differentiation, which would allow stable and specific functioning for quite a long time (from a few months to about a decade, depending on the species). This is the case with cells of the leaf parenchyma.

From cells of the leaf parenchyma, a process of cellular senescence may be analysed. First of all, on the scale of the entire leaf, an internal signal may act (from another part of the plant, for example) or an external signal. External signals especially are always present, particularly thresholds of short day and low temperature. Sometimes, these two thresholds must be present together.

Following these signals, distress hormones may intervene, particularly abscisic acid, jasmonic acid, and ethylene. Abscisic acid influences general processes; jasmonic acid, which has been studied only for a short time, affects proteases and structures of thylakoids; and ethylene amplifies phenomena of senescence that have already been triggered.

At the cellular level, increase in membrane permeability is observed: sugars, amino acids, and ions will be able to leave cells easily. Enzymes of lysosomes will be liberated (hydrolases, peroxidases, transaminases, etc.), as well as various proteases.

At the chloroplast level, the proteins of the stroma are affected first, then those of thylakoids. The case of chlorophyll-protein complexes is particularly interesting. Is it necessary to first deactivate chlorophyll, or even apoprotein? The products of degradation of apoprotein are amino acids and it would seem interesting to begin with them, but chlorophyll must not remain in the free state. Free chlorophyll cannot transmit its energy to suitable acceptors (see Chapter 2) and the energy thus serves to form single oxygen, which causes a premature destruction of cell components. Thus, chlorophyll is destroyed on its proteic support and inactivated as soon as possible, by action of chlorophyllase and other enzymes (Fig. 4.8). Once tetrapyrrol is open, proteolysis occurs. Cell senescence leads to leaf senescence.

Senescence of an entire organ—leaf, flower, and fruit, in particular—is found to result from that of all the cells. Specifically, a transitory

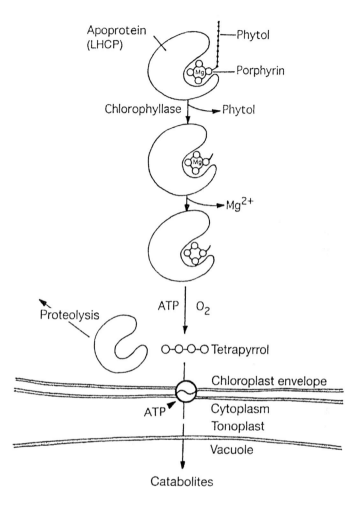

Apoprotein (LHCP)

Phytol

Porphyrin

Chlorophyllase — Phytol

Mg²⁺

Proteolysis

ATP | O₂

O-O-O-O Tetrapyrrol

Chloroplast envelope

ATP | Cytoplasm

Tonoplast

Vacuole

Catabolites

The primary intervention of chlorophyllase is general (in higher plants, algae), and it is normally followed by the removal of Mg^{++}. The amino acids can be recovered by means of an ATPasic transporter.

Fig. 4.8. Degradation of chlorophyll-protein complexes during senescence of leaves
(Matile, 1991, in Larcher, 1995).

increase in respiration is observed: **climacteric respiration** resulting from extramitochondrial oxidation and peroxidation. We also see a massive export of soluble sugars and amino acids. Then, on the basis of a signal, and following cellular senescence, there is a senescence proper to the leaf and fruit, which is abscission. A tissue of abscission forms that causes the leaf or fruit to fall.

At the level of the entire plant we observe coordinated senescence of leaves. In *Arabidopsis thaliana*, a therophyte, senescence of the rosette leaves is first observed, which begins when the plant has begun to flower, followed by senescence of the leaves of the principal stem and finally leaves of the secondary branches (Blaise et al., 1998).

In other therophytes, such as those that develop after a rain in the deserts, senescence will be global, affecting all the leaves together. In the geophytes, senescence quickly affects the entire aerial part. In many herbaceous dicotyledons, or even in tropical phanerophytes, leaf senescence is often observed from the base (the first leaves to have appeared) towards the tip (most recent leaves), on a leafy branch.

The senescence of an entire plant, or of part of a tree, is also organized, with the intervention of internal or external factors. In a perennial monocarpic plant, it is an internal factor (something that comes from mature fruits, or a disequilibrium that comes from this maturation) that triggers the senescence of the whole plant and its death. It can be prevented by removing the flowers or even the fruits if they are not yet mature.

In some trees that grow very old and in biotopes with a difficult climate, for example *Pinus aristata* or even *Juniperus utahensis*, often an entire part (several large branches, or several trunks) is observed to be completely dead (Figs. 4.9, 4.10). This may be a result of lightning or periods of climatic stress (cold and/or drought in these trees of the semi-desert), or the tree may be programmed to allow a portion of itself to die in order to preserve the remainder.

In conclusion, many interesting aspects of senescence remain to be studied, with respect to internal signals, from the cellular scale to the entire plant, as well as variations of the proteic dynamics, and finally the cycling of products resulting from degradation in the context of the general economy of the plant.

5. SEASONAL AND MULTI-ANNUAL VARIATIONS OF DEVELOPMENT

5.1. Phenology

Very early on, people understood the utility of observing certain species and recording the time of seedling emergence, leaf expansion, flowering, fruiting, and leaf fall. Many also understood the importance of climatic variations in these observations from one year to another. Farmers relied on traditional knowledge and on what they observed in the field when sowing and reaping. Since the Middle Ages, the date of flowering of cherry trees has been regularly observed in Japan. In Europe, particularly in England and Scotland, since the 18th century, observations have been

This tree of 5 to 6 m height is probably 500 years old.

Fig. 4.9. An old tree of the semi-desert (*Juniperus utahensis*)
(*photo by J.C. Leclerc*).

It is more than 4000 years old and still reproduces.

Fig. 4.10. One of the oldest trees on earth (*Pinus aristata*)
(*photo by J.C. Leclerc*).

regularly recorded for several species. A study by Menzel and Fabian (1999) shows an increase in the vegetative period in Europe, which is on average 10.8 days, between 1959 and 1991.

Some simple reference points such as beginning of leaf expansion, beginning of flowering, and date of fruit maturation may vary to some extent from year to year (climatic effects). They may vary between geographic sites, or at the same time under different microclimates. Some of these variations may be explained by the effect of thresholds of temperature or humidity, for example. Global warming also has an impact.

5.1.1. Phenology on the annual scale

It is interesting to consider not only the visible stages of development—phenophases—but also histological changes (appearance of secondary xylem, for example) or enzymatic changes (appearance of oxidases, etc.), which, taken together, may indicate the more or less hidden characters of the environment.

Phenological diagrams can be traced for different species, or even for a single species according to different places, or for ecotypes living in neighbouring places (Blaise et al., 1998). Isophanes can be traced for a single species in a large geographic area, for example, to show the date of flowering of lilacs in Europe (Fig. 4.11). Another example is fruit maturation observed in a particular region as a function of altitude (Fig. 4.12).

In temperate zones, a clear influence of temperature thresholds is noted that has given rise to the system of degree-days, which is used by biogeographers to study the extent of certain types of vegetation and can also be used in phenology to study the maturation of a fruit tree, for example.

Precipitation as well as latitude may also play a significant role (Fig. 4.11), in relation with the photoperiods. Sometimes, the conjunction of two factors is observed: for example, east of the Central Massif, a precocious yellowing of foliage (precocious by two or four weeks in comparison to the average) was observed in September 1996, caused by low temperatures and the effect of short days, which was accentuated by heavy cloud cover.

In the tropics, zones with a marked dry season must be distinguished from zones with regular rainfall, which is often the case in the equatorial zones. If there is a marked dry season, there will very often be leaf fall and fruit maturation in that season. Flowering and especially formation of new leaves and new branches will occur early in the first half of the rainy season.

In the equatorial zone, where the tropical rain forest is common, narrow fluctuations in photoperiod, and small fluctuations in rainfall with one or two relatively dry periods, affect the plants, albeit with significant variations according to the species.

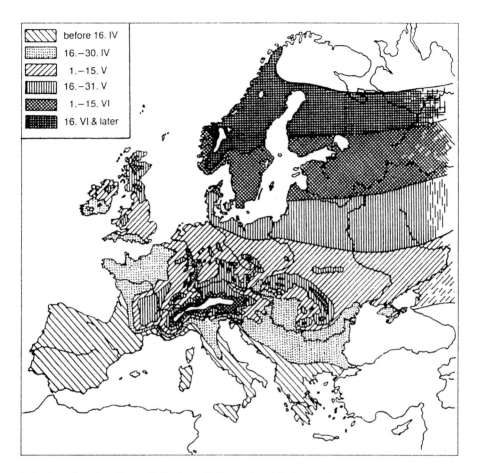

It is clear that the effect of latitude and altitude is critical. Also, in the most temperate part (France, UK), the ocean has a beneficial effect on the climate.

Fig. 4.11. Flowering of lilac as a function of geographic situation
(Larcher, 1995).

Statistics are indispensable in drawing conclusions on the overall scale of equatorial ecosystems. We can see that new leaves develop generally during periods of equinox. Plants are greatly sensitive to photoperiod; they can detect variations of some tens of minutes (low thresholds) and trigger leaf fall or flowering.

Sometimes, hydroperiodic inductions of flowering can be observed, without there being a true flowering season. A comparative study of boreal plants with high thresholds of detection of photoperiod and equatorial plants with low thresholds of detection of photoperiod would be useful.

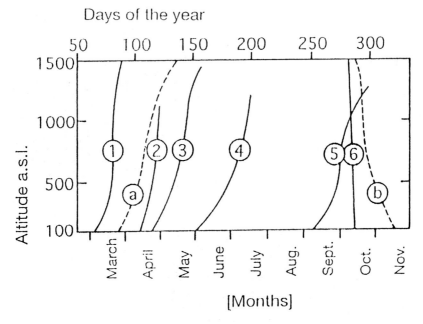

Observations were made in the Austrian Alps. 1, flowering of *Tussilago farfara*. 2, flowering of wild cherry. 3, flowering of lilac. 4, maturation of cherries. 5, maturation of cherries. 6, leaves of beech lose their green colour. In *a*, thawing of soil surface. In *b*, freezing of soil surface. In certain cases (4), it can be seen that the effect of the altitude is marked. In others (6), the photoperiod seems mainly to trigger the change.

Fig. 4.12. Effect of altitude on the phenology of plant development
(Roller, 1963, in Larcher, 1995).

5.1.2. Phenology and multi-annual variations

There are, of course, good years and bad years, which have an effect partly on the precocity of flowering or fruit maturation, but we are yet to develop an overall picture. This is where, at least in the case of phanerophytes, observation of wood growth is useful.

At the beginning of the vegetative phase, the cambium becomes active again and forms spring wood. If the preceding year was good and favourable conditions of temperature, light, and rainfall prevailed for a long time, the spring wood formed during the current year is thick. When conditions become partly unfavourable (low temperatures, aridity), cambial cell division is less frequent and autumn wood forms, often much thinner and darker than the spring wood.

By definition, the size of annual rings is a phenological character that represents overall the favourable or mostly unfavourable nature of a year. It is also possible to measure other, more refined characters such as thickness of cell walls, ratio between fibre volume and vessel volume, or size of ligneous rays to better assess the various environmental factors.

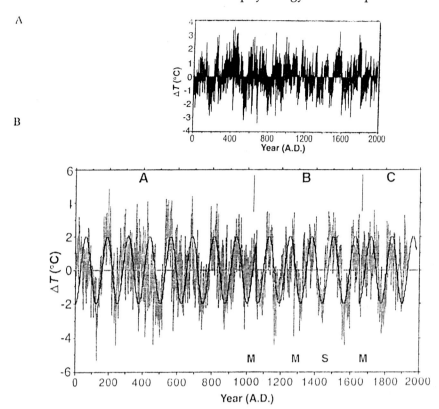

At the outset, the basis is 2000 years of measurement of annual rings on *Pinus balfouriana* (foxtail pine). Part A represents the fluctuations in average temperature measured from June to the beginning of January. A cyclic rhythm is suspected. Temperatures from 1951 to 1970 were 0.12°C above the general average. Part B represents a mathematical transformation. The temperature differences are calculated over 50 years. The last point represents the difference between the temperature in 1980 and that in 1930. A periodicity of 125 years is noted, but with two marked anomalies towards 1050 and 1650, which correspond roughly to the period of the small ice age.

Fig. 4.13. Dendrochronology and mathematical reconstruction
(*Scuderi, 1993*).

From fluctuations of these annual growth rings, the tree's life can be reconstructed: dry periods sometimes lasting several consecutive years, cold periods, or periods that were too overcast. We also see changes in the microenvironment, for example, the fall of a large tree nearby, which improved the illumination and thus the growth of the tree being studied, or, on the other hand, the rapid growth of a neighbour of another species that marked the beginning of a shaded, or even polluted, period. By examining several trees (Fig. 4.13), if possible several species, of known behaviour, using dendrochronology, we can ultimately reconstruct the history of the climate over a few centuries or even, in some cases, a few millennia.

Chapter 5

Plants under Stress

It can be said that in extreme biotopes a permanent situation of stress prevails, but we can see that this is debatable. Even in environments in which plant growth is exuberant, stresses are found that are not due to the biotope, but linked to the biocoenosis: e.g., stress due to high density, stress faced by plants of the undergrowth when the overhanging canopy captures all the available light, or stress caused by pathogens or predators that find favourable conditions. There are more or less permanent stresses or, at least, frequent stresses. There are also really exceptional situations that occur in environments that are "usually" quite favourable, for example "the flood of the century", fire caused by a storm, the "drought of the century", or even a very late spring thaw. These are the most typical stress situations in that they last a short time, occur suddenly, and are unpredictable. In the first section of this chapter, stress is defined and its major characteristics are described.

1. DEFINITION

The concept of stress can be said to imply, on the one hand, a more or less abrupt deviation from normal (average) conditions of the plant or animal and, on the other hand, a sensitive reaction of the individual in various aspects of its physiology, which change measurably with either adaptations to the new situations or, in the worst case, degradation with a fatal result. We thus have two aspects of the concept of stress: one is the **external constraint** and the other is its result on the individual, the response of stress or **state of stress**. We can find an analogy in physics, for example, in the exertion of a force (the constraint) on a coiled spring which bounces back, the rebound representing the response of the spring. We must therefore clearly distinguish the stress factor or constraint from the state of stress (response), which follows more or less quickly and is followed to some extent by an adaptation.

In a particular context either the constraint aspect or the physiological aspect may be emphasized. In studies by Levitt (1980), an attempt was made to understand how a constraint acts in structures, in what manner it produces its specific effects, up to the level of the cell cytoplasm (stimulus-oriented approach). Only constraints that have effects up to the cytoplasmic level, and thus produce a cellular reaction, are considered. In effect, the plant has mechanisms that can hasten, retard, or diffuse the action of a constraint. There is stress *avoidance*, but if the constraint reaches the cellular contents, those contents will react and show *tolerance* and then *adaptation*. Of course, one seeks to analyse the specificity of active factors bringing about immediate effects of stress, followed by the reaction and acquisition of a mode of resistance. On the basis of researches at the molecular level, such an analysis is drawn through ecophysiological methods from the cell to the plant community.

Sometimes, as in the works of Selye (1936, 1973), emphasis is laid on analysis of physiological reactions leading to the overall response of an organism. The analysis of the overall response allows us particularly to take into account physiological effects that are not specific to a precise nature of stress. Selye highlighted the notion of **stress syndrome**, which involves the equilibrium that appears between the process of destruction (distress) that is the direct effect of constraints on cells and the process of adaptation (eustress), which tends to avoid a lethal result and to restore the organism's original state (homeostasis) or something close to it, in any case a **stability**, sometimes at the cost of an exaggerated or overprotective response. It is the equilibrium that determines whether the plant thrives, or thrives partly (see Fig. 4.9), or barely survives, or dies (Fig. 5.1). The result thus depends on the equilibrium varying in value and over time, as we will see in this chapter.

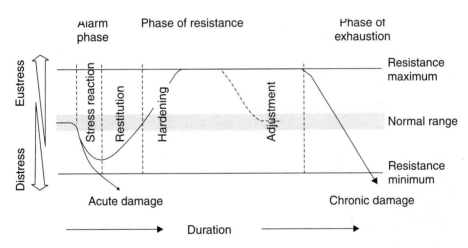

The succession of phases results from short- or long-term adaptations as well as the immediate equilibrium between phenomena of stimulation (eustress) and phenomena of inhibition (distress) of vital activities. Note particularly the alarm phase, restitution phase (rapid attempt at counter-attack), hardening phase (which takes a long time), which may be followed by a phase of exhaustion, or a return (adjustment) to a normal state of activity, more or less close to that in the beginning.

Fig. 5.1. Successive phases of a stress
(Selye, 1936, 1973, Stocker, 1947, in Larcher, 1995).

2. WHAT HAPPENS DURING STRESS?

2.1. A sequence of events

From the dynamic perspective of Selye, a succession of characteristic phases can be observed depending on forces of eustress and distress (Fig. 5.1).

When a constraint reaches the cellular level, the *alarm phase* begins immediately. As when an army launches a surprise attack on a country, the immediate result is not favourable. In the case of cells, the alarm phase begins with the destabilization of some structures, particularly membranes and proteins, and some functions. Catabolism prevails over anabolism. Very rapidly, repair processes begin and the initial state is restored. There is a synthesis of protection molecules, and overall the anabolism becomes greater than catabolism. This is the *recovery reaction*. The plant thus returns to the initial state and, if the stress factor continues, or even intensifies, the plant takes precautions and accentuates its protection processes. The plant then moves to the next phase, in which certain characters will be greater than in the average or normal state. This is the *resistance phase*, manifested particularly in plants by *hardening*, which is an adaptation. During this adaptation (understood broadly), we can distinguish an induction time, a favourable period of development of the plant under attack, with threshold effects and effectors, as well as external signs showing the adaptation by its simple recovery or by a form of hardening (Table 5.1). The resistance phase may be very long. However, if the stress factor either intensifies greatly or continues for a long time, there is a *phase of exhaustion* with appearance of major damage due to the factor itself or to the attack of predators or parasites that take advantage of the weakness of the plant, which often has a fatal result. The succession of phases is not the same in all cases (Fig. 5.2).

Major complications are the duration of stress, which may influence the exhaustion of energy reserves, for example, and the intensity of stress, which may vary with time. The interaction of these factors causes a change in metabolic equilibrium and thus the general state or external appearance of the plant. The studies of Amzallag and Lerner (1995) on this aspect are fundamental (Tables 5.1, 5.2, 5.3).

2.2. Evaluation of stress

Stress is evaluated on the basis of its manifestations. Taking into account the variety of physiological characters in different species, a state of stress can be detected only by **comparison** with the state considered normal.

The state of stress can be detected, of course, by visible signs (in the broad sense) of physiological destabilization particularly during the reaction of stress in the alarm stage, as well as in the evaluation of

Table 5.1. Induction of acclimatization (adaptation in the broad sense) to stress by pretreatment in sublethal conditions (Amzallag and Lerner, 1995)

Stress	Induction (days)	Developmental window	Effect	Remarks
Sorghum bicolor NaCl	21	Yes	Tolerance increases	Medium must be stable, effect depends on cultivar
Mesembryanthemum crystallinum NaCl	20	No	Transformation from C3 to C4 metabolism	Induction threshold at 100 mM
Hedisarum carnosum NaCl	NI	NI	Tolerance increases	
Zea mays Flooding	4	NI	Formation of aerenchyma	The effector is ethylene
Helianthus annuus Flooding	NI	NI	Formation of aerenchyma	Induction of cellulase
Holcus lanatus Cadmium	10	NI	Root growth increases	
Betula verrucosa Nitrogen deficiency	20	NI	Adaptation	In light, the leaves yellow. After adaptation, leaves become green
Lolium perenne Nitrogen deficiency	NI	NI	Plants adapt their pumping capacity to the existing nitrate concentration	For adapted plants, the relative growth rate is independent of nitrate concentration
Alocasia macrorrhiza Light intensity	15-20	Yes	Anatomical and physiological adaptations	
Fragaria virginiana Light intensity	7-13	Yes	Anatomical and physiological adaptations	
Several species Hardening to cold	3-30	NI	Acclimatization	Sensitivity to photoperiod
Medicago sativa Hardening to cold	NI	NI	Acclimatization	Acclimatization depends on cultivar
Cucumis sativus Pathogens	6	Yes	Resistance induced	Signal for systemic resistance comes from leaves contaminated locally for 3 days, systemic resistance in 6 days
Nicotiana tabacum TMV and pathogens	12	Yes	Induction of systemic resistance	Resistance is non-specific
Sinapsis alba Drought	12	NI	Many initiations of lateral roots	Several other species studied by N. Vartanian's team

Induction is time required for induction of acclimatization. Development window is period after which adaptive development occurs. NI, effect not indicated.

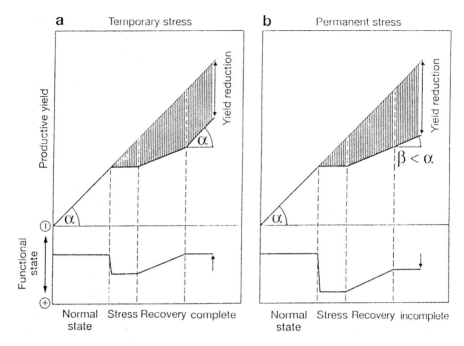

Two cases are presented. At left, after the alarm phase and recovery phase, there is no resistance or exhaustion strictly speaking, and the initial state is reached again. At right, there is probably an equivalent of partial exhaustion and productivity is reduced.

Fig. 5.2. Major phases of plant reaction to stress and the effect on production
(Hartel, 1976, in Larcher, 1995).

processes of repair, resistance, and protection. These constitute the criteria of stress.

The specific factors (or constraints) of stress require a precise and specific target in the plant. For example, intense light acts in certain conditions by directly causing damage in the thylakoids. Soluble aluminium acts by deteriorating the surface of young roots (De Lima and Copeland, 1994). In each case of a stress factor, **specific symptoms** are observed. Subsequently, characters of resistance and adaptation appear in the plant that may also be specific, such as activation of certain parts of the genome and, thus, synthesis of new messenger RNA, stress proteins, or isoenzymes. For example, in a C3 or CAM plant such as *Mesembryanthemum crystallinum*, saline stress causes the appearance of CAM (Amzallag and Lerner, 1995). In terms of resistance, we also see protection, the appearance of structures or molecules that are not specific to a single stress but to several stresses, such as changes in peroxidase activity, accumulation of anti-oxidants (ascorbic acid, tocopherol, carotenoids with several double links), osmotically active substances (polyols, proline, betains), secondary metabolites intervening in many defence processes

Table 5.2. Effect of exogenous hormones on stress adaptation of whole plants, organs, or cell cultures (Amzallag and Lerner, 1995)

Stress	Hormone	Effect
Sorghum bicolor	ABA	accelerates adaptation
NaCl	CK	inhibits adaptation
	GA	disturbs adaptation
Tobacco cells	ABA	accelerates adaptation
NaCl		
Medicago sativa	ABA	ABA and short days increase acclimatization
Cold	GA	GA and long days decrease acclimatization
Bromus inermis cells	ABA	stimulates hardening to cold
Cold		
Solanum commersonii	ABA	stimulates hardening to cold
Cold		
Detached leaves of tobacco, wheat, barley	CK	stimulates hardening to cold
Cold		
Pisum sativum	CK	stimulates hardening to cold, but only from
Cold		October to April
Acer negundo	GA	inhibits hardening to cold
Cold		
Medicago sativa	ABA	stimulates hardening to cold
Cold	GA	inhibits hardening to cold
Zea mays	Ethylene	stimulates adaptation by favouring
Anoxia		formation of aerenchyma
	ABA	stimulates tolerance
Phaseolus vulgaris	ABA	stimulates resistance
Pathogens		

ABA, abscisic acid. CK, cytokinins. GA, gibberellic acid.

Table 5.3. Effects of environmental stress on quantity or modification of cellular DNA (Amzallag and Lerner, 1995)

Stress	Adaptation	Type of DNA affected
Linum		
Mineral nutrition	NI	rDNA increases or decreases
Cold	Yes	rDNA increases
Species of Cruciferae		
Cold	Yes	rDNA increases
Zea mays		
Ultraviolet rays	NI	increase in movement of transposable elements
Hordeum vulgare		
Aluminium	NI	total DNA increases
Lobularia maritima		
NaCl	Yes	total DNA increases
Hordeum vulgare		
NaCl	NI	rDNA decreases
Solanum tuberosum		
Cut wound	Yes	total DNA increases

NI, not indicated. rDNA, repeat DNA (heterochromatin).

(polyphenols, anthocyans, various flavonoids), and so-called distress hormones (abscisic acid, jasmonic acid, ethylene) (Table 5.2). Physiological changes are observed, such as increase in respiration, changes in membrane potential and transport, and decrease in fertility, and, generally, modifications in DNA (Table 5.3). Among criteria for detection of a stress state, particular emphasis should be laid on the very great and broadbased increase in energy demand. This increase results in a *drop in the energy charge*, which falls below 0.6 (Fig. 5.3). High ATPase activity can be

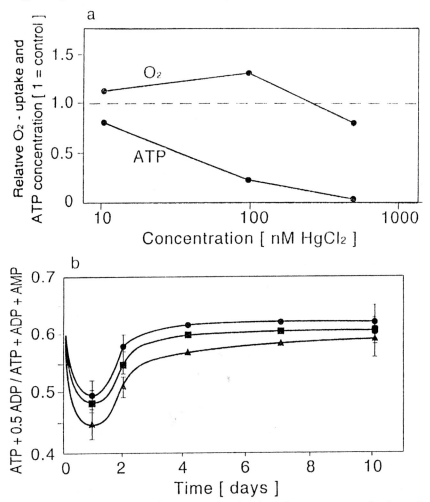

On the Y-axis, energy charge. On the X-axis, duration of measurements in days after application of sublethal concentrations of 50 μm $ZnCl_2$ (circles), 0.1 μm $CdCl_2$ (squares), or 0.01 μm $HgCl_2$ (triangles).

Fig. 5.3. Effect of stress caused by toxic salts on the energy charge of a unicellular alga
(*De Filippis et al., 1981, in Larcher, 1995*).

observed, for example, in halophytes, which must accumulate in their vacuoles Cl⁻ and Na⁺ ions reaching the cytoplasm passively by the apoplastic route. In this case, a "constant" stress prevails.

Stress factors can be detected from the reaction of particularly sensitive plants, which are *bioindicators*. The reaction of *in vitro* cell cultures can also be studied, or cultures of algae that constitute *biomonitors* and possibly are involved in the realization of *biocaptors*.

3. THE EFFECTS OF STRESS AND ITS MANAGEMENT

3.1. Diversity, rhythmicity, risks, degrees of response during the plant's life

Generally, the behaviour and dynamics of any living organism are stimulated by the intervention of *moderate stress* during its lifetime.

In plants, as in animals, stress is not only a constraint but also a stimulant, which will definitively promote the appearance of genotypes better adapted to a certain type of stress, whether it is abiotic (difficult environments) or biotic (an easy living environment leading to concurrences and competition). The intensity as well as duration of stress influences the plant dynamics. There is a switch from sexual reproductive strategies to vegetative propagation, and in the long term from strategies of reproduction to strategies of survival at the individual and species level (conservation). Stress may also cause the formation of new species. Stress often affects the life of the plant as a whole, and the plant may retain a memory of previous acclimatizations (Table 5.4). Often, the exact causes of stress are poorly defined. When a plant or tree is simply transplanted from one climate into an apparently identical one, it may show a sign of distress. For example, Yersin attempted to acclimatize cinchona in Indochina (Bernard, 1955). The plants indicated, by flowering at an

Table 5.4. Parameters of germination of sorghum *(Sorghum bicolor* L. Moench) from parents adapted or not adapted to salinity (Amzallag, 1996)

	C1	R1	A1
Fresh weight of seeds (g)	5.02 ± 1.93 a	5.61 ± 2.49 ab	6.25 ± 3.00 b
Stem/root ratio	1.46 ± 0.12 a	1.48 ± 0.11 a	1.59 ± 0.16 b
Adventitious root/seminal root ratio	0.35 ± 0.29 a	0.32 ± 0.17 a	0.45 ± 0.8 b
Seed weight at start (mg)	35.6 ± 6.3 a	27.6 ± 10.8 b	20.9 ± 10.0 c
K level in above-ground parts (meg/l)	153.1 ± 8.5 a	153.0 ± 9.1 a	145 ± 10.2 b

All the seeds grew in a non-saline environment. C1, germination from parents not subjected to salinity. R1, germination from parents subjected to salinity late (and in which only basic resistance was evident). A1, germination from parents subjected to salinity from the 8th day and having acquired adaptation to salinity. Plants of the second generation were harvested 18 days after germination. The means and standard deviations were taken from 30 samples. If the index letters are the same for two consecutive values in the same row, the difference between the two populations was not significant at P = 0.05 (*t* test).

unusually young age, a "will" to salvage the species. The stress triggered a certain number of mediators on the part of the plant: ABA in case of drought (Fig. 3.12) and other hormones on cellular models as well as on entire plants (Table 5.2).

Stress is ultimately normal for a plant, not exceptional. It helps extend the processes of selection, expanding the capacities of phenotypic variations of plants and thus favouring the appropriateness of species to their environments, which will never be constant.

To consider the subject from a more ecophysiological angle, we must examine the cost of managing stress, the multiplicity of interacting stress factors, and the consequent physiology, in order to outline the solutions that plants discover.

3.2. Cost of stress management

In a stress situation, the importance of destruction is analysed from the whole plant to the cell. The loss of structures and molecules (hydrolysis, oxidation, etc.) results in a significant loss of energy in the form of heat. Also, there is a need to eliminate or neutralize undesirable molecules and to synthesize necessary molecules. Stress is therefore expensive. The plant may, however, react and demonstrate, for example, a thermal adaptation (Fig. 5.4). The overall situation must be managed.

A plant that always lives in a hostile environment adopts a survival strategy. It cannot produce a large quantity of biomass, or a good energy yield. The energy losses are great and the plant must compromise between production and survival. Often, the plant is small, grows slowly, has reduced aerial organs, and is poorly lignified (since too much energy is required to produce lignin). In a situation of biotic aggression, what is consumed by predators must be replaced, and consumption must be limited. The plant must synthesize defence molecules, many of which, such as terpenes and terpenoids, are also costly (Table 5.5).

In a seasonal stress situation, such as very cold winters, the plant must harden, protect itself at various levels, and organize itself to that end. Production must be brought to a halt, and it can start up again in spring only progressively, with a gradual lifting of the hardened condition (Fig. 5.5). This shortens the duration for which the plant can produce at normal rates within the more or less short favourable season.

3.3. Interactions of factors and physiology

In nature, plants face rougher, and certainly more complex, conditions than in a laboratory. For example, a light stress is often linked to a drought stress and to a cold or heat stress, and sometimes to wind stress. There is thus is a complex overall stress, the constituents of which may compensate for each other or, more often, work in *synergy*, so that the stress is intensified to more than the sum of the two (or more) stress

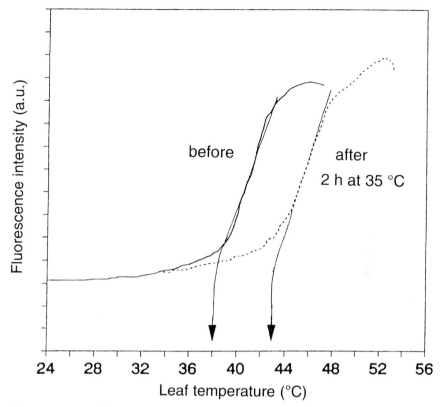

When the yield of fluorescence in low excitation light (Kautsky effect) rises, it indicates deterioration of photosynthesis (disorganization of antennae, blockage of centres, etc.) Deterioration appears at a temperature just 5°C higher if the leaves have been adapted to heat (2 h at 35°C). The test was carried out by progressively increasing the temperature of samples (1°C/min.) while measuring fluorescence.

Fig. 5.4. Yield of fluorescence and thermal adaptation of potato leaves
(Havaux, 1993).

factors. This applies, for example, to the joint effect of cold and intense light (Fig. 5.6). On the other hand, the effect of a stress seems to provoke a reaction of the plant permitting it to resist other stresses. Resistance to cold as well as drought is observed, and that resistance proves to be progressive, for example, in *Pinus cembra* in the early winter. A saline stress applied to dolichos bean leads to the extension of resistance to cold and heat.

The means of protection are based chiefly on the adaptation of membrane structures and on cytoplasmic stabilization through synthesis of special proteins, which provide a joint protection. The proteins of heat, drought, and osmotic shock seem, if not identical, at least similar; during a thermal shock, for example, there is synthesis of *flavonoids*, which also

Table 5.5. Cost of reconstruction and defence (Larcher, 1995)

Substance or organ	g glucose/g matter
Defence substances	
Tannins	1.55 to 1.6
Cyanogenic glucosides	1.9 to 2.1
Alkaloids	2.8 to 3.3
Latex	3.3
Wall substances	
Lignin in conifers	2.44 to 2.49
Lignin in angiosperms	2.48 to 2.52
Cost of replacement of organs	
Soft leaves rich in defence substances	1.8
Soft leaves poor in defence substances	1.3
Sclerified leaves	1.35 to 1.55
Conifer needles	1.5
Non-lignified branches	1.1 to 1.35
Lignified branches	1.4 to 1.55

Either to replace what has been consumed by herbivores or to synthesize defence substances (in g of glucose equivalents), to reconstruct 1 g of dry matter in general or 1 g of a specific substance. Certain values are high not only because of the energy value, but also because of many successive reactions of synthesis.

protect the plant against browsing, ultraviolet rays (Fig. 5.7), and parasitic fungi.

With respect to the different resistances, we must ask whether, certain stresses being seasonal or even daily, plants are capable of anticipating and preventing attacks that they "know" will occur at a particular time through mechanisms of endogenous circadian rhythms or ultradian rhythms (most often annual).

3.4. What solutions does the plant find?

Whether the plant is under stress continually, seasonally, or occasionally, it must survive. Survival for the plant means escape from or avoidance of stress, with resistance, and recovery or restoration, if avoidance is not possible.

Escape or avoidance may be spatial, as with geophytes and hemicryptophytes in winter, which stay underground or very close to the soil. Or it may be temporal, as with spring therophytes that avoid growing during the cold season or hot season.

If escape or avoidance is impossible, the plant must face and resist the stress, which it can do only within certain limits. A colony of red algae (Leclerc, 1976) with an optimal water potential of the medium at 0.5 to 1.5 MPa can resist an abrupt increase in salinity of about –1 MPa, but not of –2 MPa, which leads to cell death.

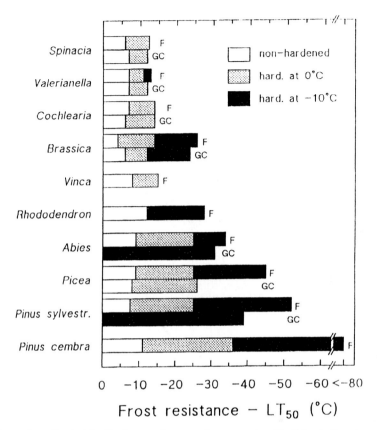

Resistance is estimated by the temperature causing necrosis of half the leaf surface (LT_{50}). The white bars represent the non-hardened plants, the grey bars represent plants hardening at 0°C, and the black bars represent plants hardening at –10°C. Plants were hardened either in the field (F) or artificially in a growth chamber (GC).

Fig. 5.5. Frost resistance of higher plants as a function of conditions and hardening
(Bauer et al., 1994).

The development of **resistance** is particularly well studied in the case of resistance to cold, i.e., **hardening**. Hardening is one of the most fascinating phenomena at the cell level and is of enormous economic significance, particularly in forestry and fruit arboriculture. Resistance in the form of hardening to cold is manifested at the cellular level by a drop in water potential. This drop is parallel to a decrease in water content, with formation of less ice and thus a lesser contraction of cells.

Some of the substances that cause the drop in water potential also have directly anti-freeze properties. In freezing conditions, these cause either a superfusion or a gradual transformation from the liquid state to the solid state by a great increase in viscosity and synthesis of structures (Fig. 5.8), with formation of glucosides and special proteins. On the extracellular scale,

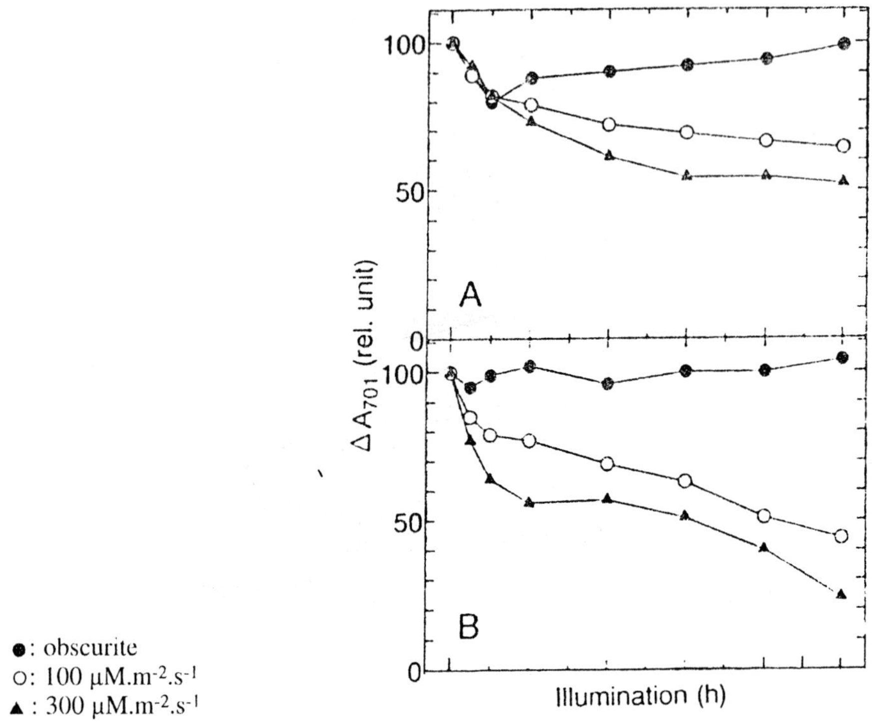

●: obscurite
o: 100 µM.m⁻².s⁻¹
▲ : 300 µM.m⁻².s⁻¹

Closed circles, dark. Open circles, 100 µM/m²/s. Triangles, 300 µM/m²/s. The thylakoids were treated for 15 min. to 5 h in fairly low light and at either 5°C (part A) or 25°C (part B). The effect was measured (Y-axis) by the spectrophotometric dose of P_{700} subsisting. The effect of inhibition of photosystem I is still more marked at 25°C than at 5°C. There is probably a biochemical effect associated with the photochemical action itself. It should be noted that the thylakoids suffer from an illumination of 100 µM/m²/s, at 5°C as well as at 25°C. The entire leaf is, in physiological conditions, insensitive to photoinhibition at intense light and at 25°C. Sonoike believes that, during the isolation of thylakoids, a protective factor disappears that is effective at normal temperature.

Fig. 5.6. Rate of development of photoinhibition of photosystem I, in isolated thylakoids
(Sonoike, 1995).

because of the low water potential, less ice forms in the apoplastic domains. Also, isolated structures are observed, for example, at the buds, which in effect slow the rate of freezing in the cellular structures.

After resistance, there must be total or partial **recovery**. With red algae, after a shock of –1 MPa, a decline in photosynthesis and the growth rate is observed during the alarm phase, but after a few days the algae recover so far as to increase photosynthesis and growth, although not to the initial level.

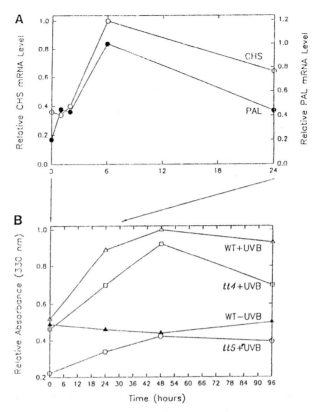

The *Arabidopsis* plants were cultivated without UV-B for the first 10 days following germination (0 h corresponds to the end of this period), then UV-B rays were added (8 Kj/m^2/d) or not added. In A, the relative contents at 1 for the best conditions are represented. Black circles represent mRNA level of phenyl ammonia lyase (PAL). White circles represent mRNA level of chalcone synthase (CHS). In B, relative absorbance at 330 nm of leaf extracts from plants raised in 16 h day and 8 h night was measured, with or without UV supplement. WT + UVB: wild colony treated with UV after 10 d, white triangles. WT − UVB: no UV, black triangles. Mutant tt4 + UVB: white squares. Mutant tt5 + UVB, white circles.

Fig. 5.7. Sensitivity of mutant *Arabidopsis* to UV-B at the level of flavonoids
(Chen, et al., 1993).

In the case of saline stress, salts more or less "invade" the plant at different cells and organs (Table 5.6). An internal management appears, showing an adaptive distribution and segregation of ions. The effects are observed not only at the individual level, but over several generations. Saline stress of parents has been observed to influence the behaviour of offspring with respect to salinity (Tables 5.4, 5.6, 5.7).

On the scale of the entire plant, part of the individual may die, although the whole survives by means of underground parts, as during the cold season or following a brushfire, then restores itself, regenerating from the salvaged parts. Also, there are overall changes in reproductive

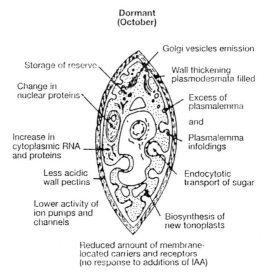

The cambium, in view of its size, must resist frost particularly well. Note particularly the import of sugars and increase in cytoplasmic proteins, as well as wall thickening, and reduction of the vacuole size.

Fig. 5.8. A cambial cell becoming dormant and hardened in October
(Lachaud, 1989, in Larcher, 1995).

Table 5.6. Distribution of Cl⁻ and Na⁺ ions in halophytes in different organs of
Halimione portucaloides in a natural station (Leclerc, 1964) and in cellular
compartments of leaves of Suaeda maritima cultivated on a substrate
with 350 ml/m³ NaCl (Flowers, 1985)

Organ or cellular compartment	Proportion of cellular volume or water content of organ (% H_2O/fresh matter)	Concentrations (mol/m³)		Cl/Na
		Na⁺	Cl⁻	
Leaves	86	440	414	0.94
Large branches	64	252	229	0.91
Large roots	77	192	209	0.92
Small roots	70	417	317	0.76
Cytoplasm	11	116	60	0.52
Chloroplasts	1.4	104	98	0.94
Vacuole	81.6	494	352	0.71
Walls	6	194	138	0.71

strategies, the floral apparatus being particularly fragile. Such changes are seen in plants of hostile environments (high altitudes, for example), which reproduce by vegetative means either normally or seasonally.

The short- or long-term consequences of stress on growth can be modelled (Stuart-Chapin III et al., 1993). Since the 19th century, natural constraints have been distinguished from those due to the presence of

Table 5.7. Effect of late exposure to salinity on sorghum seedlings from parents adapted
or not adapted to salinity (Amzallag, 1994)

	NaCl in medium, mol/m^3			
Lot	0	75	150	225
C1	6.81 ± 1.11 a	5.86 ± 0.87 b	5.17 ± 1.33 c	2.51 ± 0.51 d
A1	6.21 ± 1.41 a	5.25 ± 1.44 b	5.88 ± 1.83 ab	3.89 ± 1.12 c
R1	5.90 ± 2.43 a	5.05 ± 1.57 a	4.86 ± 1.59 a	2.48 ± 0.60 b

The dry mass of aerial parts was measured. The plants were exposed to different levels of
salinity 21 days after germination till harvest on day 34. Twelve plants were measured for
each treatment. C1, germination from parents not subjected to salinity. R1, germination from
parents subjected to salinity late (in which only basic resistance was evident). A1, germination from parents subjected to salinity from the 8th day and having acquired adaptation to
salinity. Statistical data are the same as for Table 5.4.

humans (anthropogenic constraints) in the aggravation of stress effects
and the reactions that follow.

4. STRESS IN PRACTICE

4.1. Introduction

In this section we focus on the effects of light, drought, cold, heat, salinity,
external characters of soil, and then constraints due to the biocoenosis and
human constraints, some of which have led to new stresses or amplified
existing stresses.

4.2. Constraints and responses

Natural constraints can be classified in two major groups. The first group
comprises constraints of the biotope: intense or inadequate light, heat or
cold, excessive humidity or dryness, and salinity. These constraints do not
correspond to general environmental factors, such as light, temperature,
water potential, pH, and so on, in themselves, but to their **extremes**. The
second group consists of constraints arising from other living things:
pathogenic organisms, predators, or neighbours.

The constraints of human origin affecting the plant world include
what is generally called atmospheric pollution—CO_2, SO_2, fluorine,
various aerosols—and pollution of soil and water by aluminium, acidity,
heavy metals, and other factors. We must also include agriculture and
cultivation practices. Often, human constraints are complex and cover
many factors, and some of them amplify constraints that were originally
natural. There are constraints that comprise natural and man-made
factors.

Many human constraints have a major impact not only on individuals, but also on the evolution of species.

4.3. Constraints and responses due to the biotope

4.3.1. Illumination

Illumination stress can be of two types: either the illumination is too little, or it is too intense (Fig. 5.9), as when a tree falls and plants of the undergrowth are suddenly exposed to the sun. Or illumination may be of normal intensity but accompanied by cold or drought, so that the plant cannot use the light and the light becomes dangerous to the plant by way of phenomena of photosensitivity.

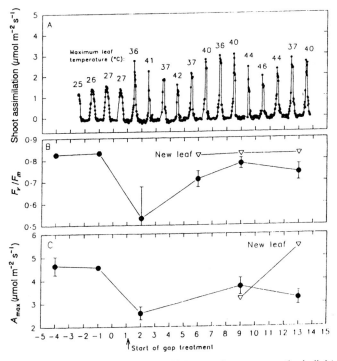

The plant studied was *Alocasia macrorrhiza*, which grew for many months in light attenuated by screens (around 0.9 mol·photons/m²/d instead of 24 mol·photons/m²/d at discovery). The effect of a gap in the canopy was simulated for 13 days by illumination for 3 h/d with halogen lamps (1900 μmol/m²/s). The maximum daily temperature of an apparently intact leaf was measured. Part A is drawn from measurements of overall net photosynthesis of the plant, from the assimilating leaf area, i.e., without taking into account the leaf area that was necrosed in the first hours of treatment. After 3 days of decline, photosynthesis did not resume in the surviving leaves. To these surviving leaves were added new leaves, which showed better adaptation after B and C. The maximal rate of photosynthesis (in saturated light and optimal temperature), after an abrupt fall, was only partly restored in old leaves that survived. Also, when the days were very hot (44 to 46°C from the 9th to the 11th day of simulation), overall assimilation declined. In fluorimetry, Fv/Fm seemed to reflect accurately the maximum photosynthesis of part C.

Fig. 5.9. Adaptation of leaves of a bush of the undergrowth, in a simulated situation of wood fall *(Mulkey and Pearcy, 1992)*.

The major aspects of the problematics of photosynthesis and environment are presented in Chapter 2. Some of the effects of water deficit experienced at mid-day (under intense light) normally lead to the intervention of the xanthophyll cycle. A stress situation occurs when the external constraints exceed the normal capacities of the plant to manage an excess of light energy (via synthesis of proteins, xanthophyll cycle).

In a temperate environment, when illumination becomes too intense after a tree has fallen or been cut down, certain species of the undergrowth, which cannot reach a stage of recovery and restoration, disappear. *Oxalis acetosella* plants, for example, whiten in a few days and rapidly disappear. In the mountains, especially, very young plantations of fir (*Abies alba*) or spruce (*Picea abies*), in the months that follow tree felling and fresh plantation, show signs of photoinhibition, particularly the loss of chlorophyll by photosensitization, to a greater or lesser extent depending on the individual. This leads to the death of part of the population.

In a tropical environment, the situation is probably more complex, with many species showing a great adaptive capacity. Refined, non-destructive studies can be made using the techniques of induction of chlorophyllian fluorescence that were described in Chapter 2. For example, a study conducted in Panama (Thiele et al., 1998) showed that, following a tree fall, plants of the undergrowth received direct sunlight for 1 to 2 h, and fairly severe photoinhibition was produced in several species studied, which is a stress situation. As the light became less intense, the plants recovered gradually, coming back to the initial state by early evening, but the recovery began only when the stress ceased.

During photoinhibition, some of the symptoms represent photooxidation, which results from the appearance of highly reactive oxygen forms (Table 5.8). This follows, at least partly, photosensitivity that activates the excitation energy of chlorophyll not to liberate an electron, but to form triplet chlorophyll (see Chapter 2). The triplet chlorophyll, in

Table 5.8. Reactive forms derived from molecular oxygen (Millet, 1992)

O_2^-	Anion superoxide (also a radical)
HO_2	Radical perhydroxyl
H_2O_2	Hydrogen peroxide
HO	Radical hydroxyl (the most oxidant form)
$^1O_2$	Single oxygen (potentially even more dangerous than HO)
ROO	Radical organic peroxy
ROOH	Organic hydroperoxide
RO	Radical alkoxyl
RO*	Excited carbonyl

The oxygen molecule is a good oxidant (+0.82 V), but from the kinetic point of view it is poorly reactive since at physiological temperature it requires a catalyst to act. The anion superoxide is the source of the other reactive forms.

turn, gives energy to a molecule of O_2 (if the plant does not have enough carotenoids to capture all the energy from the triplet chlorophyll), which is transformed into the particularly destructive singlet oxygen.

The effects of photoinhibition vary according to the O_2 level, CO_2 level, and temperature. Although oxygen can show a harmful effect, some experiments show that photorespiration can reduce photoinhibition (Epron, 1997), but is this possibly because O_2 is used at the level of chloroplasts and thus will be at a low quantity? Low levels of CO_2 favour photoinhibition, particularly at high temperature.

There are often interacting effects of photoinhibition by intense light and drought. Photosynthesis may be considerably slowed and, thus, a large quantity of incident light energy is excessive, with the possibility of forming a great deal of singlet oxygen. A good means of defence (Fig. 5.10) is the formation of a large amount of violaxanthin in the framework of the xanthophyll cycle, which can practically be detected by a reflectance index, but is not always efficient at high altitudes (Lutz, 1996). In the longer term, the plant can also synthesize coloured secondary metabolites that serve as a filter (Fig. 5.11). This synthesis is stimulated by intense light and UV rays.

The impact of cold also favours photoinhibition but it may be an indirect effect (Lutz, 1996). A low temperature greatly slows biochemical reactions. If a plant under intense light shows photosynthesis limited by cold, the chlorophyll of the centres liberates energy with difficulty in the form of electrons. Singlet oxygen may also form in varying amounts. This leads to photooxidation of lipids and proteins. In this case, another indirect effect of cold is, for example, that when the rate of proteic synthesis is slowed, the synthesis does not compensate for the photodestruction. Thus, light stress must be studied not alone but in association with another aggravating factor.

Overall, the plant makes integrated responses to different environmental stresses (Fig. 5.12). Those responses often lead to a slowdown in growth.

4.3.2. Hydric stress

As with light, two extreme situations may occur more or less frequently and regularly in the life of a plant. There may be too little water, or too much. The plant reacts, for example, by exposing fewer leaves to the sun and thus to heat (Fig. 5.13) or by protecting it with trichomes (Fig. 5.14).

Drought stress has always been of interest to farmers and ecophysiologists, but research has intensified in recent years. Even in temperate zones, water has become expensive and worth economizing on. In some areas, in the context of the present climatic evolution, periods of drought seem to have increased.

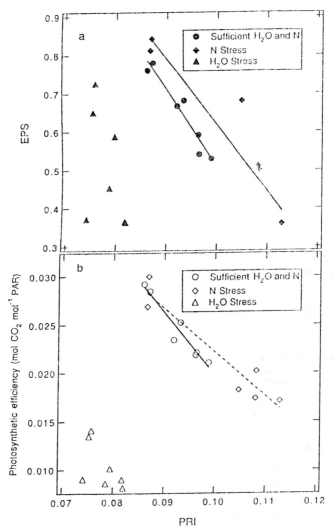

Fig. 5.10. Reflectance, photosynthetic activity and xanthophylls
(Gamon et al., 1992, in Schulze and Caldwell, 1995).

Gamon et al. previously observed that in the region 500-550 m significant variations of reflectance could be observed as a function of the physiological state. More particularly, a close correlation exists between reflectance at 531 nm and the state of epoxidation, a physiological index of vegetation: PRI = $(R_{ref} - R_{531})/ (R_{ref} + R_{531})$ was defined, where R_{ref} is a reflectance of reference (fairly unvarying) estimated as a function of the species and equal to 550 nm in sunflower. This index is compared to a state of epoxidation, EPS = $(V + 1/2A)/(V + A + Z)$. V = violaxanthin, A = antheraxanthin, and Z = zeaxanthin. In control sunflowers and nitrogen-deficient sunflowers, correlations of PRI and photosynthetic efficiency on the one hand and PRI and EPS on the other are close enough. In contrast, in plants that are drought-stressed, correlations are low. One explanation may be that the state of epoxidation is a means of defence with a high rate of violaxanthin that may be found in plants that show only a low photosynthetic activity.

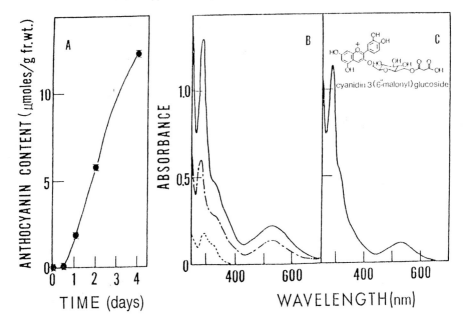

Accumulation of anthocyans was studied in artichoke cell cultures, stressed by UV light. (A) There is accumulation after a latency period (left). (B) There are progressive changes in the overall absorption spectrum at the beginning (dotted lines), and there are only carotenoids. (C) The absorption spectrum of an anthocyan is shown with its significant peak in near UV.

Fig. 5.11. Secondary metabolites and protection against UV and excess visible light (*Kakegawa et al., 1991*).

4.3.2.1. Drought stress

Drought stress occurs in a zone in which the availability of water corresponds to values of water potential that are between the loss of turgor (causing a visible but recoverable wilting) and the point of permanent wilting. For most plants, those values are around −1 to 2 MPa.

In Chapter 4, we saw that there is a gradient of water potential in the plant. If the leaves are still turgid, they may have a water potential that is lower than that of the roots, while the latter are close to plasmolysis.

In a soil in which the water potential has dropped and become close to what has caused plasmolysis at the root level, water leaves the root instead of being absorbed by it. It is a stress situation and the plant must react, at two levels. A signal, perhaps at the membrane level of absorbent hairs, indicates to the root cells that the soil water potential has become insufficient. The root **reacts,** the reaction being increasingly evident as time passes, by emitting a secondary signal (an internal messenger) in the form of release of ABA (Table 5.9). This internal messenger, probably transported by xylem (see Fig. 3.4), brings about another reaction at the leaf level. The

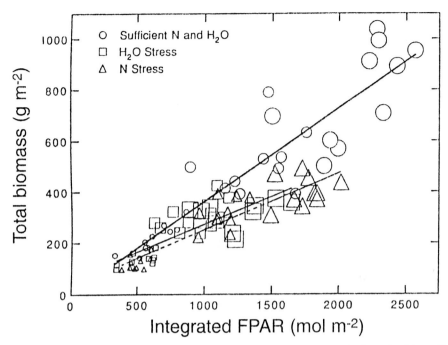

Measurements were taken in sunflower fields. On the Y-axis, integrated light (over time) in mol/
m². On the X-axis, biomass obtained after harvest of material at different periods of growth.
The symbols are larger for data at the end of the season. Circles: good nitrogen and hydric
uptake. Squares: hydric stress. Triangles: nitrogen stress.

Fig. 5.12. Relation between biomass and various stresses
(G. Joel, in Schulze and Caldwell, 1995).

leaves still hold a great deal of water, but that reserve is threatened by the
fact that transpiration will no longer be compensated by root absorption
of water. The leaves react by closing their stomata under the influence of
ABA. The plant will thus maintain its hydric state efficiently depending
on how effective the cuticle barrier to transpiration is.

The cuticle barrier constitutes a long-term reaction, or an adaptation
acquired over evolution. If the barrier is ineffective, the water will
gradually escape, and if the soil remains at a very low water potential, the
plant will in fact exhaust itself and inevitably wither. This happens, for
example, with plants of the undergrowth such as ficaire, which has a low
cuticular resistance.

If, on the contrary, the barrier is effective, as it frequently is in many
Mediterranean heliophytes such as garden geraniums, the plant will hold
water for a long time. It will resist and put out defensive mechanisms, for
example by synthesizing anthocyans and carotenes that in the absence of
photosynthesis (CO_2 uptake being nearly nil) will allow the plant to filter
light and to extinguish single oxygen molecules.

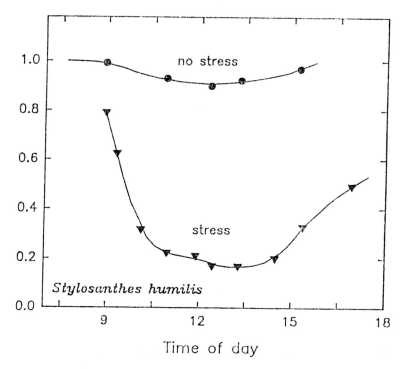

Stylosanthes humilis can direct its leaves towards the light (cosine of the angle of incidence is close to 1) perpendicularly to the rays, if it can utilize all the energy received. In good hydration (−0.2 MPa), the cosine is observed to remain close to 1. On the other hand, in the stressed plant (−2.6 MPa at mid-day), the leaf is sufficiently well oriented only at the start of the day, because of the nocturnal hydration and the relatively low temperature. At mid-day the cosine falls to 0.2 and the plant absorbs only 15 to 20% of the light energy in relation to the possible maximum. By this means, it limits the effects of photoinhibition. The great hydric deficit having considerably reduced CO_2 uptake and enzyme activity, the flow of electrons that can leave the photosystems is very much reduced.

Fig. 5.13. Hydric stress and leaf orientation
(Begg and Torsell, 1974, in Schulze and Caldwell, 1995).

The plant, which conserves water and thus its metabolic capacity, can in the process of its **resistance** reduce its leaf water potential by means of, for example, hydrolysis of polysaccharides, which will enable the increase of suction pressure and allow the plant to resume absorption of water through the roots. One may go so far as to say that there is **recovery**. Quite often, in fact, there will be a point of equilibrium between the foliar and root suction (the root water potential) and the soil water potential, which will keep water from leaving or entering the plant for fairly long periods in the xerophytes.

This process of reaction, resistance, and recovery is illustrated in the following examples.

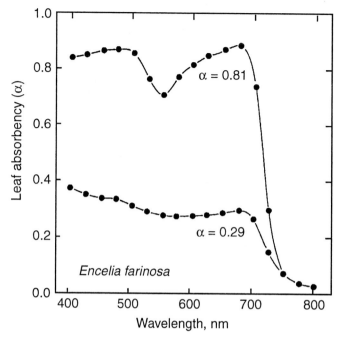

The power to absorb light (α) is greatly reduced (by 3 times) in very pubescent leaves. It is no longer possible to distinguish the depressions of absorption in green, probably because the residual light that crosses the trichome layer becomes quite oblique and therefore has a long optic trajectory facilitating its absorption.

Fig. 5.14. Protective role of leaf trichomes
(Ehleringer and Bjorkman, 1978, in Schulze, 1995).

Hydric stress is always linked closely to photosynthesis and to its effect on transpiration. The water potential at which stress occurs varies greatly according to the species. First of all, tolerance of desiccation is observed to differ widely among the various taxonomic groups and biological types (Table 5.10). Plants considered to be inferior may be highly resistant. Cormophytes are often sensitive and may be either stenohydric, i.e., able to support only limited variations in potential, which applies to most species, or poikilohydric, i.e., capable of tolerating much greater variations. Note also that ferns, although commonly under-stood to be closely associated with humidity, are quite resistant.

Because of their adaptation to the environment, plants are able to live habitually, and even during the unfavourable season, in a situation in which they are clearly short of the water potential (or water activity) that is dangerous for them in that it leads to stress that may be fatal for all or part of the plant (Table 5.11). Plants of the temperate oceanic environ-ments are particularly very far from the values of water activity that will cause problems.

Table 5.9. Effects of low water potential on primary roots of maize
(Saab et al., 1992, in Munns and Sharpp, 1993)

	Rate of elongation		Final cell length		Rate of cellular production	
	(mm/h)	(% of control)	(μm)	(% of control)	(cell/h)	(% of control)
Control	2.81	100	260	100	10.8	100
Low	0.92	33	120	46	7.6	70
Low + FLU	0.27	10	59	23	(4.6)	(43)

FLU: fluoridone 10^{-5} M. The experiment was conducted at -1.6 MPa (low) in the presence or absence of an inhibitor (FLU) of ABA accumulation. The cellular elongation was measured on cortical cells. Measurements were taken 48 h after the transplantation of seedlings in the treated medium.

Table 5.10. Tolerance of desiccation (data from many authors including
Gaff, 1980, in Larcher, 1995)

Plant group	Relative humidity of air (%)	Corresponding water potential (MPa)
Marine algae		
deep water	97 to 99	-4 to -1.4
shallow water	86 to 95	-20 to -7
rocking marshy zones	83 to 86	-25 to -20
Hepaticopsidae		
hygrophytes	90 to 95	-14 to -7
mesophytes	50 to 92	-94 to -11
xerophytes	0 to 36	$-\infty$ to -140
Mosses		
hygrophytes	90 to 95	-14 to -7
mesophytes	50 to 90 (10)	-93 to -14 (-310)
xerophytes	5 to 10	$-\infty$ to -400
Hymenophyllaceae	75 to 90	-38 to -14
Prothallus of ferns		
Pteridium aquilinum	> 90	up to -14
Cystopteris fragilis	> 90	up to -14
Asplenium ruta-muraria	40 to 60	-120 to -70
Tissues of stenohydric cormophytes		
epiderms	96	-6
mesophyll	95	-7
root cortex	97	-4
Poikilohydric cormophytes		
Borya nitida	85 to 90	-22 to -14
Xerophyta villosa	66	-56
Myrothamnus flabellifolia		
in natural site	11	-298
in laboratory	0 to 11	-218 to ∞

According to assays done on cells for 12 to 48 h, in chamber by measurement of minimum relative humidity of air, not leading to damage.

Table 5.11. Hydric stress and index of relative water deficit
(most data from Larcher, 1973, Bannister, 1976, Sveshnikova, 1979, and
Bobrovskaya, 1985, in Larcher, 1995)

Type of vegetation	Region	Ratios observed often or ( ) exceptionally
Scrub	Mediterranean	
	northern limit	40 to 70
	southern limit	80 to 85
Small shrubs	Mediterranean	90 to 105
Chamaephytes	Atlantic moors	10 to 15 (88)
Deciduous trees and bushes	Northern Europe	10 to 40 (50)
Herbaceous plants of undergrowth	Northern Europe	6 to 25 (50)
Dry fields	Central Europe	40 to 85
Mesophile fields	Central Europe	20 to 40 (75)
Low prairies	Central Europe	50 to 90 (108)
Low steppes	Central Asia	60 to 80
Desert plants		
Trees and bushes	Karakorum	30 to 50
Chamaephytes and large grasses	Central Asia	50 to 65

An index of relative water deficit can be defined as the ratio of deficit (observed habitually in the most unfavourable part of the season) of water vapour saturation over the deficit of water vapour saturation that effectively causes measurable damage in a plant (Hofler et al., 1941).

When controlled stress is imposed in an experimental situation and the consequent reactions studied, generally a correlation is observed between stomatal conductance and photosynthesis (Fig. 5.15). However, in many situations of hydric stress, it is seen that photosynthesis is affected by drought when stomatal regulation does not occur or occurs only partly. The "stomatal reaction" of the plant is not always sufficient.

Generally, photosynthesis declines during the morning (Fig. 5.16) and the activity of water diminishes. Stomatal regulation sets in rather late and ensures a stable compromise between transpiration and photosynthesis, both being relatively restricted. This strategy subsists in a stress situation, except when the stress is too intense, in which case, if it is a plant that tolerates drought well, it closes its stomata entirely and thus is effectively "dead" in terms of photosynthesis.

Very often, in case of experimental stress (Fig. 5.17), the stomata close quickly and more or less completely, depending on the species. The weeping birch (Betula occidentalis), which prefers moist environments, reacts slowly, closing half its stomata. However, Betula pubescens, which prefers a drier environment, such as a mining site, completely closes its stomata in an hour.

The kinetics of reaction are ultimately very different. For example, the elder reacts quite slowly but strongly. Often, plants of the undergrowth

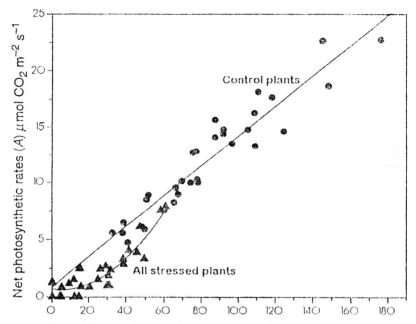

A very good correlation can be seen between photosynthesis of control olive trees and stomatal conductance (Y-axis). On the other hand, the relationship is more complex in stressed plants in which photosynthesis may be very low while stomatal conductance is not reduced. All the results of three levels of stress are combined in the graph.

Fig. 5.15. Correlation between net photosynthesis at a given moment
and stomatal conductance
(*Angelopoulos et al., 1996*).

react quickly, for example, *Aegopodium podagraria* (Fig. 5.17), which closes its stomata in half an hour. Other plants of the undergrowth are even quicker. *Alliaria officinalis* closes its stomata in 7 to 10 minutes, and *Ranunculus ficaria* in 4 to 8 minutes, but these plants have a low cuticular resistance.

Resistance to hydric stress depends not only on the species and the stress value, but also on the development stage the individual is in or on particular conditions. It also depends reciprocally on acclimatization. Moreover, the plant's adaptation to drought affects its capacity to resist other stresses.

In oat (Table 5.12), a light stress may be applied in early development (equivalent to a dry spell in spring). Significant effects are observed in the vegetative apparatus, but there are also later effects seen in the reproductive apparatus. If a more severe stress is applied during pollination (equivalent to a serious drought in the beginning of summer), we see that the vegetative apparatus is overall not affected more than with the spring stress, but the effect on seed production is much more severe.

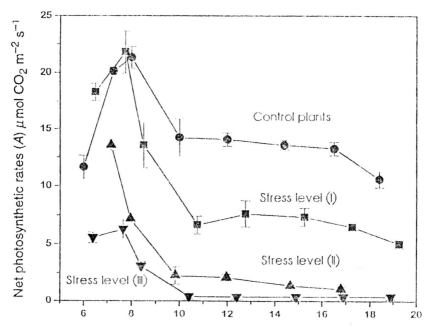

The study involved 2-year-old olive trees cultivated in a pot and planted in the field. Measurements were taken in the field in July, in southern Italy. In the control plants, where soil water was maintained at around 85% of field capacity, leaf potential varied from 0.45 MPa at daybreak to −2 MPa towards noon. When stomatal regulation occurred, the potential was restored slowly. The watering was restricted to varying extents to subject the plants to three levels of stress, over 8 to 10 days. At stress level I, leaf potential was −1.2 MPa at daybreak, at stress level II it was −4.25 MPa, and at stress level III it was −5.7 MPa. At level III, leaf potential dropped to −7 MPa towards mid-day and did not rise again during the day. A drop in photosynthesis was always observed during the morning, but around 10 to 12 h the stomatal regulation allowed maintenance of a certain level of photosynthesis, which would be very low once the stress became severe (level II).

Fig. 5.16. Daily variation in photosynthesis under controlled water stress
(Angelopoulos et al., 1996).

Different species will react in diverse ways to water stress (Fig. 5.18). The reaction varies as a function of the real drought, but also as a function of other environmental factors such as CO_2 and O_2 levels (Fig. 5.19). For example, it is noted that the resistance of sunflower to water stress is quite good in low oxygen, but that of maize is mediocre under the same conditions. If one studies the cold-drought interaction, for example in maize (Table 5.13), one sees that in the absence of acclimatization to drought, the recovery is downright poor after cold stress. However, if there is acclimatization, the recovery is better, albeit not complete. The stomatal regulation is restored (according to values of C_i CO_2), but the water use efficiency and photosynthesis are reduced, perhaps because of an incomplete recovery at the metabolic level.

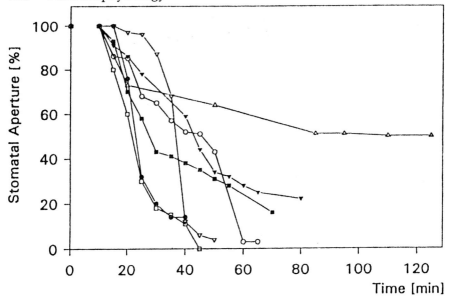

• Aegopodium podagraria   ○ Betula pubescens
▾ Atropa belladonna   ▪ Rosa spec.
▫ Avena sativa   ▽ Sambucus nigra
△ Betula occidentalis

Opening of the stomata is indicated in percentage of the opening measured before the drop in water potential.

Fig. 5.17. Kinetics of stomatal closure under hydric stress
(Kappen et al., in Schulze and Caldwell, 1995).

The mediator of reactions to drought is ABA. The presence of ABA can be demonstrated in the xylem of a stressed plant (Fig. 3.4), and we also see that ABA induces a diminution of leaf growth (a relatively long-term reaction), which adds to the rapid, classic effect on closure of stomata.

Abscisic acid also allows direct adaptation at the root level in case of hydric stress (Table 5.9). In maize, if ABA accumulation is inhibited, it can be seen that the growth of roots is slowed down much more than in the presence of the inhibitor, which implies that ABA will also keep the plant from suffering an excessive reaction.

The situation is ultimately quite complex, with a great many interactions, as we will see later, and the subject is still being researched extensively.

4.3.2.2. Inundation stress

Excess of water is also a source of stress. In some cases, a site is well-drained: water circulates in the soil and is often oxygenated. In others, the

Table 5.12. Oat subjected to precocious or late drought stress: components of yield (Peltunen-Sainio and Makela, 1995)

Character	Fairly light stress at beginning of development							
	Control		Stressed		Ratio		Significance	
	Mean	CV (%)	Mean	CV (%)	(%) S/T	G	T	G × T
Total mass (g/plant)	5.28	29	1.47	29	57	***	***	**
Vegetative mass (g/plant)	1.33	28	0.84	28	63	***	***	*
Panicle weight	1.25	34	0.63	40	50	***	***	***
No. filled grains	29	40	16	51	55	***	***	***
No. empty grains	19	39	13	53	68	***	***	***
Grain wt (mg)	43.1	32	39.4	55	91	***	ns	***
Harvest index (%)	48	14	42	24	88	***	***	***

More severe stress during pollination stage							
Control		Stressed		Ratio		Significance	
Mean (%)	CV	Mean (%)	CV (%)	S/T	G	T	G × T
2.05	32	1.34	23	65	***	***	***
1.13	29	0.95	22	84	***	***	ns
0.92	53	0.40	34	43	***	***	***
18	79	2	169	11	***	***	***
21	49	23	36	110	***	***	***
51.1	128	20.0	173	39	**	ns	***
42	29	30	17	71	***	***	***

The tests of significant different were done between genotypes (G), treatments (T), and genotype-treatment interactions (G ´ T), ***$P \leq 0.001$, ** $P \leq 0.01$, * $P \leq 0.05$, ns = not significant. CV %: coefficient of variation. The grains marked empty were in fact either empty or incompletely filled.

water does not move and inundation stress is often linked to anoxia stress. General studies have been devoted to this particularly important subject in the case of certain cultivated fields and natural riverside forests (Siebel et al., 1998).

In particular, we find adaptations of the vegetative apparatus and physiological adaptations at the level of root activity. In certain cultures, the seedlings are especially affected, as in soybean (Zhang and Davies, 1987), but also perennial plants (olivella et al, 2000) at the bormonal and physiological responses. From the vegetative point of view, we can cite the capacity for stem elongation in, for example, African rice plants of the Niger basin. In temperate zones, many bushes and shrubs of riverside forests have flexible stems, as does the highly fibrous currant bush.

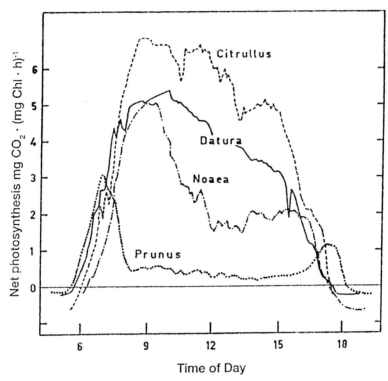

Several behavioural regimes were exposed: *Citrullus colocynthis* and *Datura metel* were watered and suffered only a little at the end of the day, while the water vapour gradient became steep because of high heat. *Noaea mucronata*, which is a desert species, was not watered and greatly reduced photosynthesis by late morning. *Prunus armeniaca* suffered greatly from the lack of water and did not photosynthesize as well as at the beginning of the day.

Fig. 5.18. Daily variations of photosynthesis of plants of the semi-desert, at the end of a long dry season *(Schulze et al., 1972, in Larcher, 1995).*

4.3.3. Thermal stress

4.3.3.1. Heat stress and adaptation

Adaptation of living things to heat has a long history, probably linked to the considerable greenhouse effect and volcanic activity that prevailed in the early periods of the biosphere.

If we limit our discussion to plants taken in the broader sense, i.e., including bacteria and fungi, we observe considerable adaptations in the "lower organisms" (see Chapter 1), particularly in the archaebacteria, and in resistance forms in the eubacteria. It is useful to refer in this context to a fundamental work (Pelmont, 1995).

In the tracheophytes (Table 5.14), we see that thresholds of sensitivity to heat are clearly higher than those found habitually in natural

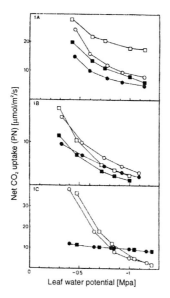

Uptake (PN) was measured as a function of leaf water potential, in the presence of different levels of O_2 and different light intensities. (A) Sunflower leaves. (B) Bean leaves. (C) Maize leaves. Closed circles represent the results under 21% O_2 and 250 μmol photons/m²/s PPFD, and open circles represent 21% O_2 and 850 μmol photons/m²/s PPFD. Closed squares represent the results under 2.5% O_2 and 250 μmol photons/m²/s PPFD, and open squares represent 2.5% O_2 and 850 μmol photons/m²/s PPFD. In low oxygen the resistance of sunflower to hydric stress is quite good, unlike that of maize.

Fig. 5.19. Net CO_2 uptake as a function of water potential, O_2, and light *(Scheuermann* et al., *1991).*

Table 5.13. Acclimatization to drought mitigates the effects of cold in maize
(Irigoyen et al., 1996)

Treatment	Net CO_2 assimilation rate (μmol CO_2/m²/s)	Intracellular CO_2 concentration (μl/l)	Water use efficiency (μmol CO_2/mmol H_2O)
5 days of cold			
control	23 ± 0.2 a	240 ± 7 a	2.1 ± 0.2 a
treated	0.8 ± 0.1 b	301 ± 5 b	1.3 ± 0.1 b
Drought + cold	1.2 ± 0.1 b	256 ± 8 c	2.6 ± 0.4 a
Recovery after 5 days			
control	3.9 ± 0.4 a	166 ± 9 a	3.6 ± 0.4 a
treated	1.9 ± 0.3 b	258 ± 10 b	1.4 ± 0.2 b
Drought + cold	2.5 ± 0.2 c	212 ± 16 a	2.1 ± 0.2 c

A sensitive variety of maize (cv Errazo), cultivated in the north of Navarre (Spain) in an Atlantic climate, was used. The effect of cold (5°C) was measured at the end of the 5th day. Subsequently, recovery was measured 5 days after the end of the treatment. The control remained always at 25°C. The experiment was conducted in a temperature-controlled chamber, under 150 μmol photons/m²/s PPFD, as were measurements of photosynthesis. Drought + cold indicates plants previously exposed to drought (17 days without watering). Different index letters indicate significant differences (P < 0.05, Student's *t* test).

Table 5.14. Temperature limits of leaves of vascular plants in different climatic zones (Larcher, 1973; Kappen, 1981; Sakai and Larcher, 1987; Nobel, 1988; additional data from Bannister and Smith, 1983; Losch and Kappen, 1983; Larcher et al., 1989; Yoshie, 1989)

Groups and zones	Sensitivity threshold in hardened state	Sensitivity threshold in growing period
Tropics		
Trees	–2 to +5	45 to 55
Plants of the undergrowth	–3 to +5	45 to 48
Plants of high altitude	–15 (sometimes –20) to –5	to 45
Subtropical zones		
Woody evergreens	–12 to –8	50 to 60
Buds of deciduous trees	–15 to –10	
Palms	–14 to –5	55 to 60
Crassulescean plants	–10 (sometimes –20) to –5	58 to 57
C4 herbs	–5 (sometimes –8) to –1	60 to 64
Desert therophytes	–10 to –6	50 to 55
Temperate zone		
Woody evergreens of oceanic zones in mild winters	–12 to –8	50 to 60
Old woody species		
Leaves	–25 to –20	45 to 50
Buds	–35 to –25	to 50
Herbaceae of sunny habitats	–20 (sometimes –30) to –10	47 to 52
Shade plants	–20 (sometimes –30) to –10	40 to 45
Grasses of the steppe	–196 to –30	60 to 65
Halophytes	–20 to v10	
Crassulescean plants	–25 to –10	sometimes 42, 55 to 62
Aquatic plants	–12 to –5	38 to 44
Stenohydric ferns	–40 to –10	46 to 48
Regions with cold winters		
Conifers	–90 to –40	44 to 50
Deciduous plants of the taiga	–196 to –30	42 to 50
Bushes of the tundra or high altitude	–70 to –30	48 to 54
Herbaceous plants of the tundra or mountains	–196 to –30	44 to 54

Thresholds of sensitivity were measured with 50% damage and 2 h exposure to cold, or 30 min. in heat (in Larcher, 1995).

conditions. Aquatic plants (which live in a thermally stable environment) and many ferns are relatively sensitive. It is the same with tropical plants, which seems curious but we must remember that tropical environments have little variation in temperature. On the other hand, some steppe grasses of the temperate or subtropical zone, which are C3 or C4 plants, are quite resistant, even more so than some xerophytes.

Heat stress is often parallel to light stress and drought stress. Its overall effects can be studied with phenomena of recovery or exhaustion leading to partial necrosis or the death of the plant. We must also, on a finer scale, look at the effects on physiology (Table 5.15).

In a single environment, two species may show clear differences in their response to heat stress. Photosynthesis is more thermosensitive than

Table 5.15. Thresholds of temperature affecting physiological reactions
(Bjorkman et al., according to Berry and Raison, 1981)

Function	Atriplex	Tidestromia
Leaf level		
Net photosynthesis	43	51
Respiration in dark	50	55
Semipermeability	52	56
Chloroplast level		
Photosystem I	> 55	> 55
Photosystem II	42	49
RUBP carboxylase	49	56
PEP carboxylase	48	54
3-P6A kinase	51	51
Adenylate kinase	47	49
Hexose phosphate isomerase	52	55
RU5P kinase	44	52

The data are based on 10% inhibition of reaction and treatment times of 10 min. on leaves of two C4 plants of the desert.

respiration. It is observed that photosystem I is very often more resistant (Havaux, 1996). Finally, there is a wide diversity at the enzyme level. We also find the presence of thermal shock proteins, comparable to what happens in the thermophilous bacteria.

In case of heat stress, it is overall the lower sensitivity to respiration that may lead to lethal situations. When the plant is at a temperature at which photosynthesis is inhibited and respiration is still active, it becomes quickly exhausted. We must also consider that respiration is not exclusively mitochondrial. Chlororespiration increases greatly (Lajko et al., 1997) during heat stress in simple organisms (cyanobacteria and algae) and competes directly with photosynthesis.

Recovery after heat stress is often slower than recovery after cold stress (Fig. 5.20).

4.3.3.2. Cold stress and adaptation to frost
There is a parallel between adaptations to drought and those to cold: small cell size and overall low water content. In the northern hemisphere, there is a clear seasonal effect on adaptation, particularly for high-altitude plants such as *Pinus cembra*, for which the lethal temperature is near –5°C in June but –40°C in winter (Fig. 5.21). This is the phenomenon of **hardening**. Hardening is the basis of effective resistance to cold in the winter (Fig. 5.22). It can be measured from changes in the physical state of cellular water. The most widespread manifestations of hardening are described below.

(1) The increase of dissolved substances in the vacuolar liquid causes a drop in ΨH_2O to –5 or –6 MPa in conifers of the taiga or the subalpine stage. The solutes are often oses, phosphorylate or otherwise, which

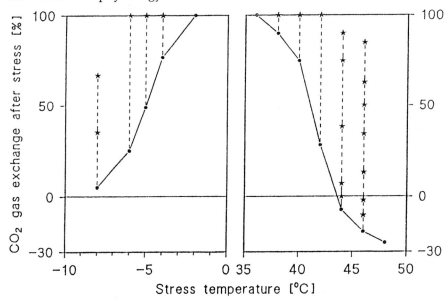

Fir needles (*Abies alba*) were studied in October (beginning of hardening period). The cold treatment (at left) lasted 12 h. After –8°C, there was practically no photosynthesis and a recovery of around 2/3 after 2 days (2*). Needles recovered in less than one day after assays with higher temperatures. The effect of heat treatment (for 30 min.) is presented at right. At 43°C, it is observed that net photosynthesis becomes negative and that recovery is not total after 5 days.

Fig. 5.20. Thermal stress and photosynthesis
(Pisek and Kemmitzen, 1968; Bauer, 1972, in Larcher, chapter 13 of Schulze and Caldwell, 1995).

derive from the hydrolysis of amidon. Part of the soluble sugars come directly from photosynthetic activity in winter. The conifers of the taiga have a clear assimilation from –6°C.

(2) Decrease in total water content. For example, the water content of rhizomes in *Poa arctica* in winter is only 36% of what it is in summer. In the taiga, in *Vaccinium vitis idea*, the water content of leaves reduces by 85% in the beginning of winter. As the water content decreases when extracellular frost begins to form, the volume reduction the cell undergoes, because of the aspiration of water caused by the formation of this frost, is mitigated. This is because the mitigation of cytoplasmic and vacuolar water loss is reinforced by the low water potential.

According to J.C. Levitt (1956), the degree of contraction $G = 50 - W_0/W_0 (C_0/C_g - 1/2\ W_0) + 50$. In this formula, W_0 is the original content of water in the cell, expressed as a percentage of fresh weight. C_0 is the concentration in dissolved substances before exposure to cold. C_g is the concentration when the plant freezes. C_0/C_g increases in plants during hardening, especially in those most resistant to cold. In potato, the "hardened" tuber freezes at –3°C and C_0/C_g will be 2.3. This is a mild

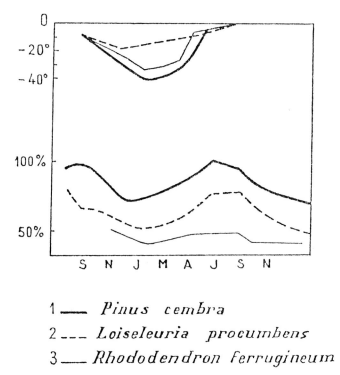

1 —— *Pinus cembra*

2 ＿＿＿ *Loiseleuria procumbens*

3 ＿＿ *Rhododendron ferrugineum*

Above, lethal temperatures. Below, water content, % of secondary weight. Note that *Loiseuleuria* is relatively non-resistant. It is a very low chamaephyte that is normally protected by snow.

Fig. 5.21. Seasonal variations of lethal temperatures and lethal threshold of dehydration in high-altitude species *(Pisek and Larcher, 1954, in Birot, 1964).*

hardening. In *Pinus cembra*, a temperature of –36°C corresponds to C_0/C_g = 10. This is a more complete hardening, which is also shown in the low value of the degree of contraction: G = 3. Also, in hardened plants an increase is observed in the fluidity of the cytoplasm.

For many deciduous trees of North America, extracellular frost in the buds appears only at about –25°C. In several plants of the tundra it appears at –30°C. The resistance increases with the low water content of buds and trunks, and an early opinion was that all the water was closely connected to the cytoplasm and the cell wall cellulose. We must consider that the hardening results from a complex set of conditions and interactions, including photoperiod, degree of differentiation, development of buds, season, and states of dormancy.

In the spring, when evapotranspiration (which can be represented as E = 4T(T + 25/20) with E in mm of water per month) becomes important again, the strong adaptation to winter cold must be accompanied by a good capacity to absorb water from soil that is still cold. Plants of the

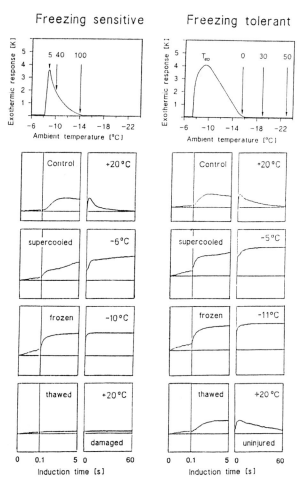

This *Rhododendron* is a native of the Alps and the Pyrenees at the alpine stage and in the upper subalpine stage. It is regularly exposed to severe cold in winter. At left are presented the results of non-hardened leaves obtained in June. At right, the leaves were hardened and measurements were taken in January. Above, the development of necrosed leaf surfaces is analysed, as well as thermal exchanges during a cooling of 1°C/min. In non-hardened leaves the exothermal peak is sharp. Freezing occurs abruptly at around −8°C and necrosis is complete at −14°C. On the other hand, in hardened leaves the exothermal peak is much more spread out. Freezing is less abrupt because of the presence of protective solutes. There is no more damage at -16°C. Fluorescence assays (3 linear time scales, continuous illumination) show that at a temperature at which water is still in superfusion (−5 or −6°C) and at a temperature at which ice forms (−10 or −11°C), fluorescence reaches a maximum from which it does not fall again. This shows that the electron transfers are soon completely blocked. When they return to 20°C, hardened leaves return to normal activity, but those measured in June practically emit no signals. The chlorophyll and other constituents are destroyed.

Fig. 5.22. Resistance to cold in *Rhododendron ferrugineum* depending on whether it is hardened or not
(Larcher, in Schulze and Caldwell, 1995).

temperate or cold regions absorb water better at about 5°C than tropical plants do. Common heather (*Calluna vulgaris*) has a root absorption of water only 20% lower at 5°C than at 20°C, but the inhibition reaches 94% in cotton (*Gossypium hirsutum*). Generally, shrubs of the tundra absorb water as well at 0°C as at 20°C.

The coldest environments for plants are not always those we might imagine. The tundra is the vegetation zone in overall the coldest climate, but the plants there are low-growing and covered with snow, so that in winter most parts of the plant have to tolerate temperatures of only −10 to −15°C.

However, here also, there are topographic situations and micro-climates in which there is no snow cover because of constant winds. *Pyrola rotundifolia* shows an absence of formation of intratissular ice up to −30°C. On particularly exposed rocks and stones in the Arctic, we find lichens that resist temperatures of −75°C, even if they are drenched, and it is usually excessive wind exposure leading to desiccation that prevents the establishment of lichens.

The most severe cold to which plants are effectively exposed is found at high altitudes and in the taiga forests for aerial parts of trees. In the Eurasian taiga, the severest cold is found east of the Ienissei River, with a west-east gradient of dominant tree species.

In eastern Scandinavia and northwestern Russia, it is mostly *Pinus sylvestris* and *Picea excelsa* that are found. The needles of these trees are resistant to temperatures as low as −38°C. Then there appear *Larix sibirica* and *Abies sibirica*. Around 53° E longitude, we find *Pinus cembra* and finally *Larix dahurica*, which are resistant to temperatures of −70°C.

In high altitudes in the tropics, the stress of cold occurs at night (Fig. 5.23) and the reactions and recovery vary according to the species (Table 5.16). In a temperate climate, and at high altitudes, as on the plains in spring, phenomena of photoinhibition may occur, followed by cold stress (Fig. 5.24).

From a physiological perspective, cold may particularly affect membrane phenomena and photosynthesis often from 0 to 10°C and, of course, cause the freezing of cells in poor conditions of crystallization when the temperature falls below 0°C. All this varies as a function of hardening (Table 5.17), which also favours photosynthesis of certain varieties of cereals. Many tropical plants undergo measurable damage between 0 and 10°C, which corresponds to the phenomenon of chilling.

Particularly with regard to photosynthesis, when we measure overall photosynthesis as a function of temperature, we observe breaks in the slope, partial activities (enzymatic activity, membrane transport, etc) show abrupt changes of energy of activation of the phenomena studied, which often implies changes in the membrane state, in terms of fluidity and transport systems.

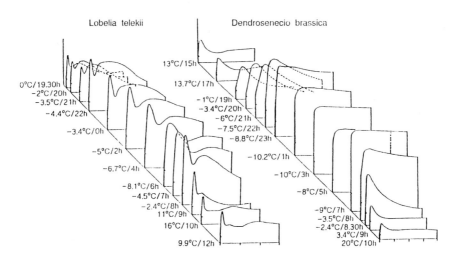

The two species, both forming large rosettes, are found in Mount Kenya at 4200 m. The circadian temperature differences are great and the night is very cold. It can be seen that both species recover well on the following day. During the night, levels of fluorescence are high, which is the sign that even in the stationary state many of the centres (*Lobelia*) or all (*Dendrosenecio*) are closed (little or no electron flow). *Lobelia* seems better adapted, even at −8°C. Part of the plastoquinone pool again becomes oxidized in the equilibrium state.

Fig. 5.23. Kinetics of induction of fluorescence measured *in situ* for 18 to 24 h
in high-altitude plants
(Bodner and Beck, 1987, revised by Larcher, in Schulze and Caldwell, 1995).

Table 5.16. Resistance to cold in tropical high-altitude plants
(Beck et al., 1982, 1984; Goldstein et al., 1985, in Larcher, 1995)

Species	Provenance	Threshold for 50% leaf damage (°C)
Espeletia atropurpurea	Andes 2850 m	−6.1
	3100 m	−8.1
Espeletia schultzii	Andes 3560 m	−10.0
	4200 m	−11.2
Dendrosenecio brassica	Mr Kenya 4100 m	to −10
Dendrosenecio keniodendron	Mt Kenya 4200 m	−14
Lobelia keleti	Mt Kenya up to 4500 m	to −20

The plants, all of large rosette form, were collected from the Venezuelan Andes and Kenya.

4.3.4. Saline stress

4.3.4.1. Presence

An abundance of dissolved salts is observed not only in a marine environment, but also in many terrestrial environments.

Sea water on average contains 480 mM/l Cl^- and other ions, including Mg^{++}, in significant quantities. The result is a conductivity of 44 mS/cm

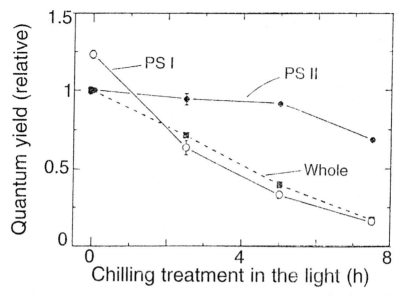

Decrease in quantum yield of non-cyclic electron flow is measured from the total photosynthesis (broken line) in the preparations of thylakoids from leaves treated for a long or short period
at 220 µmol photons/m²/d and at 4°C. Open circles: PS I electron flow. Closed circles: PS II
electron flow. Relative values at 1, which are calculated from the overall activity from intact
leaves. Note that in normal conditions it is the PS II yield that is limiting. During treatment, from
the point of 2 h 30 min. of cold, the PS I becomes limiting.

Fig. 5.24. Photoinhibition of photosystem I during chilling of cucumber leaves
(*Terashima et al., 1994*).

Table 5.17. Photosynthesis and hardening of cold-tolerant plants (Huner et al., 1993)

Species	Cultivar	$PS_{max}(H)/PS_{max}(NH)$	$Q_A(H)/Q_A(NH)$	LT_{50}
Secale cereale (rye)	Musketeer	2.30	2.20	−32
Triticum aestivum (wheat)	Kharkov	1.31	1.85	−21
	Augusta	1.23	nd	−13
	Monopol	1.03	1.50	−12
	Glenlea	0.55	1.10	−8
	Marquis	0.46	nd	−7

$PS_{max}(H)$: maximum photosynthetic capacity of hardened plants (NH: non-hardened). $Q_A(H)$:
redox state of primary acceptor of stable electrons of photosystem II ($Q_A Ox/Q_A Red$) of
hardened plants (NH: non-hardened). LT_{50}: temperature at which 50% of plants die. nd: not
determined.

and a water potential of −2.5 MPa or −25 bar. In parts high on the strand,
we may find values that are much higher.

In a terrestrial environment, there are many salt deposits in semi-
desert zones. Certain soils (solonetz) are basic, because of Na_2CO_3, and
saline. They are unsuitable for cultivation. In drought-sensitive zones,
heavy irrigation is used, but the waters that percolate into more or less

saline soils of watersheds are charged with salts. When this water is used for irrigation, it produces a relative concentration of salts because of evapotranspiration by plants. The presence of 0.2 mol/l NaCl suffices to reduce ΨH_2O from –10 bar and thus to disturb the growth of most plants. In practice, many cultivated plants are sensitive from –0.3 MPa onward, which corresponds to 4 mS/cm, and the use of water that has conductivity higher than 2 mS/cm should be avoided for irrigation.

Salt is also present in large quantities in urban environments in temperate zones, because it is widely used in winter to melt snow on the roads. This sudden and massive addition of salt results in saline stress for trees in city gardens and on the roadside. Some plants resist saline stress better than others (Table 5.18).

Table 5.18. Trees and soils that are affected by salinity or sea spray (Sucoff, 1975; Meyer, 1978; Carter, 1982; Dassler, 1991, in Larcher, 1995)

Sensitive to salt	Relatively resistant
Deciduous trees and shrubs	
Acer platanoides (Norway maple)	*Acer negundo* (North America)
Aesculus (horse chestnut tree)	*Ailanthus* (China)
Carpinus (hornbeam)	*Eleagnus*
Euonymus (spindle tree)	*Fraxinus* (ash)
Fagus (beech)	*Gleditschia* (North America)
Juglans (walnut)	*Hippophae rhamnoides* (sea buckthorn)
Ligustrum vulgare (privet)	*Lycium* (lyciet)
Platanus (plane tree)	*Potentilla fructicosa*
Prunus serotina (North America)	*Quercus* spp. (oak)
Rosa rugosa (some clones)	*Rosa rugosa* (some clones)
Syringa (lilac)	*Robinia pseudo-acacia*
Tilia (lime tree)	*Sophora* (Japan)
With persistent leaves	
Ilex (holly)	*Coccoloba*
Ligustrum lucidum	*Ficus* spp.
Mahonia (North America)	*Magnolia grandiflora*
Trachelospermum	*Nerium* (oleander)
Conifers	
Abies (fir, many species)	*Juniperus chinensis* (juniper)
Picea (spruce, many species)	*Picea mariana*
Pinus strobus (Weymouth)	*Picea parryana*
Pinus sylvestris	*Pinus halepensis*
Pseudotsuga (douglas fir)	*Pinus nigra*
Taxus (yew)	*Pinus ponderosa* (ponderosa pine)

Various authors consider leaves that have a sensitivity threshold of 0.3 to 0.5% Cl in dry weight at the beginning of summer to be sensitive. The corresponding value for conifers is 0.2 to 0.4%. Resistant conifers are those having a threshold around 0.6% and resistant leaves are those having a threshold of 0.8 to 1.6%. Many species indicated are studied because they are used in cities for decoration and are subjected to salt that is used to clear snow from highways.

4.3.4.2. Immediate consequences

To photosynthesize, a plant needs to transpire and thus to establish a gradient of ΨH_2O between the soil and the leaves. The Ψ of leaves must be lower than that of the soil. If the soil is saline, there is already a low Ψ and the plant must have an even lower ΨH_2O. To reach this very low Ψ, the plant generally resorts to accumulation of salt, which constitutes an adaptive reaction to the saline stress. The result is a series of harmful effects. The Na^+Cl^- reduces photosynthetic activity. Generally it slows enzy-matic reactions, which leads to metabolic distortions. Na^+ and Cl^- accumulate especially in the leaves, and specifically in the vacuoles (Table 4.15). Competition is also observed for the absorption of Ca^{++}, K^+, NO_3^-, and other ions. Of course, a slowing of growth is also observed.

In all cases, the plant must tolerate very low water potential, and that tolerance is similar to the response to drought stress.

4.3.4.3. Reactions

Overall, plants tolerate a drop in ΨH_2O better in the absence of salt than in its presence. In *Dunaliella* (Johnson et al., 1968; Heimer, 1973), it was clearly demonstrated that the alga accumulates large quantities of glyc-erol. At equal ΨH_2O, the enzymatic activities were much better in the presence of glycerol than in the presence of salts.

In the face of a demand for low ΨH_2O, plants thus exclude salt or absorb it, but with distribution and compensation by organic solutes. In some rare cases, there is an effective barrier against NaCl. Most often, however, it is absorbed and then eliminated by some means.

More salt accumulates in the leaves than in the stems and larger roots (Table 5.6). Specifically, salt accumulates at the vacuoles. It may accumu-late in the large vacuoles and result in succulence of the plant, as in sea beans. Accumulation is continuous, since absorption of water is continu-ous at the roots, and since a barrier to salts would cost the plant too much in terms of active transport.

Transpiration and the salt level in soils also vary seasonally. In the temperate zone, if a large amount of salt is accumulated by halophytes in summer, the accumulation becomes excessive in wetter seasons because the plant no longer needs a very low ΨH_2O, which is useless. Generally, the accumulation being more or less continuous, there must be a limit to the drop in water potential, and thus elimination by one means or another of part of the salts absorbed.

Salts may be eliminated by chemical transformation, by formation of a volatile compound, for example gaseous elimination of $ClCH_3$ by *Mesembryanthemum crystallinum*. Salts may also, partly by active transport, accumulate preferentially in the old leaves (*Aster tripolium*), which then are shed by the plant.

Often, excretion of salts is observed (mainly of chlorides) at special-ized cells in the leaves (*Tamarix*, various Plumbaginaceae, including *Limonium*, *Statice*). The salt may also be accumulated actively in the hairs, which then die, fall off, and are replaced by new hairs. As for excretion, this accumulation and elimination (in halophyte *Atriplex*, for example) requires energy that is provided by photosynthesis and is thus favoured by light.

In any state, the salt that remains in the plant is distributed, first between various organs with, for example, regulatory effects by transport in the phloem. At the cellular level, the salts being accumulated in the vacuole, there necessarily are compensatory organic solutes of water potential in the hyaloplasm and chloroplasts.

The compensatory solutes may be organic acids, oses, derivatives of oses or amino acids, sometimes ions similar to Cl^- and Na^+ but less aggressive against enzymes, such as SO_4^{--} and K^+. The vacuolar accumu-lations involve ATPase activity at the tonoplastic level. Rare organisms (prokaryotes or protists) have enzymes that tolerate NaCl and can do without compensatory substances.

In parallel with the appearance of drought stress proteins, or heat stress proteins, there are also proteins of saline stress and various changes at the level of cellular DNA (Table 5.3), accompanied by interaction with plant hormones (Table 5.2) (Claes et al., 1990). The possibilities of reaction vary with the age of the plants (Table 5.7), and they are particularly significant in the young seedlings. Young seedlings must root in the superficial soil layers, which are often rich in salts. Finally, a previous adaptation of parental plants to salt seems to have consequences for salt tolerance in their offspring.

The solutes compatible with enzymatic activity and low water poten-tial are highly varied. They include glycerol (*Dunaliella*), sorbitol, or mannitol in *Aster* and various species of mangroves, proline, or even glycine-betain in *Suaeda*, *Atriplex*, and various species of mangroves.

Some of these substances are also involved in resistance to cold or drought, which shows in part the coordinated activity of reactions of plants to stress.

Various laboratories are developing several research strategies to obtain cultivated plants adapted to saline conditions.

4.3.5. Physical events changing the limits of soils

Physical events that change the soil parameters are caving in, landslides, precipitation of volcanic ash, and anoxia. The stress on the plant is physical (breakage of trunk by rocks following a caving in, partial burial by a landslide or mudslide, or by volcanic ash) as well as chemical because, in many cases, the root system finds itself far form the new soil surface and is faced with anoxia.

A plant that is partly destroyed may eventually react favourably, as with trees and shrubs that are more or less basitonic. The individual will put out shoots at the base (e.g., birch). If it is bushy (highly basitonic), it often has a good capacity for layering, with rooting of prostrate branches. If the branches are partly buried and remain more or less vertical, adventitious roots will develop close enough to the new soil surface and help the plant avoid anoxia. The Alpine currant bush, which is very bushy and prone to layering, can thus survive in a mass of fallen rock or at the base of unstable slopes (Chiche et al., 1998). Peculiar forms may appear, such as a very thin variant of *Nigella arvensis* that was found in the island of Santorin following emissions of volcanic ash.

These land movements often lead to anoxia. The effect of anoxia may be avoided as described above, but the plant may also adapt to it directly, particularly by inhibition of alcohol dehydrogenase activities, which prevents alcoholic fermentation and toxicity by allowing the formation of lactic or malic acid.

4.3.6. Fire

Fire is a spectacular form of stress. It has a brutal, sudden, and unpredictable effect. It is often lethal. In many cases plants show avoidance. Sometimes they show a reaction of direct adaptation that may even eventually result in a positive effect, if not for an individual, at least for the future population. A brush fire rages quickly, and there is practically no heating up under the soil surface. Certain rhizomatous grasses (such as *Imperata cylindrica*) thus get their aerial parts burned but, during the subsequent rainy season, new stems grow vigorously. In such a situation, *Imperata* often grows invasively.

Some trees, such as the sequoias, have a very thick bark that is resistant to fire because it contains substances similar to tannins. Part of the oxygen realizes oxidations without heat, and thus without feeding the fire. By this means it "suffocates" the fire by the absence of a "draft". Moreover, the thick bark insulates the cambium from heat. Some plant formations find their dynamics of activity renewed and regenerated by fire, as with the fynbos of southern Africa.

Rhizomes and basal buds that are well protected survive fire. Many seeds also resist fire. In some cases (sequoia cones, for example), fire is even necessary to release the seeds and allow them to germinate (Brenckmann, 1997).

4.3.7. Wind

Wind has a major impact on the morphology of many plant formations. Certain species adapt either to a strong, often dominant wind, or to a random wind. Wind acts on the isolated tree or plant and on a group, or on an entire forest.

An isolated individual facing the wind is smaller than normal and has a thicker trunk. This is the morphology of avoidance (Fig. 5.25). For example, a hawthorn will develop a very off-centre crown, which resists the dominant wind and clearly indicates the wind direction. The same phenomenon can be observed in a group: in the Caux countryside, the farms are often surrounded by four screens of beeches. If the farm is close to the sea, we clearly see that all the crowns are more or less dwarfed and bent away from the sea, from which the dominant wind blows. In the forest, the wind results in a uniform size in the dominant trees.

An individual exposed to wind probably reacts by a mechanical signal inducing a variation of the physical effect, which causes the morphological adaptation (Nobel, 1991).

Wind acts automatically in conjunction with isolation, which favours herbivory (Aphalo and Ballare, 1995). It also causes reduction in water use efficiency by elimination of the boundary layer. The plant must therefore simultaneously synthesize defence substances against herbivores and develop thick cuticles to protect the leaves from drying out. It must make investments that compete with investments needed for construction of trunk and thick branches. Plants thus have had to evolve a compromise between these different energy costs in the face of the wind factor.

4.3.8. Pollutants

The subject of pollutants is at the forefront of effects of the biotope and of the biocoenosis, and a good proportion of pollutants are directly caused by humans.

4.3.8.1. "Natural" pollutants, human amplification of pollution and new factors of pollution

The concept of pollutant has evolved over time. Before the industrial era, certain observers studied the disagreeable aspects of the urban environment that were partly due to the high population density: putrid waters, bad air, slaughterhouses, unregulated deposit of wastes, and overcrowding of cemeteries.

Some observers in the past reported situations that appeared more or less abnormal but they were not able to distinguish those of physicochemical origin from those of biological origin (remember the concept of "spontaneous generation"!). Moreover, there were no means of measuring pollution.

With the beginning of the industrial era, there was a gradual discovery of emission into the air, water and soil of substances that existed earlier but were now found in much greater quantities. New substances were also discovered that had never before existed in nature.

Thus, the notion of pollutant emerged as representing anything that appeared new in the environment (essentially physical or chemical char-

This hawthorn lives on a chalky and windy plateau in the Connemara. Note the compact shape of the crown and the bent of the tree in the direction of the prevailing wind.

Fig. 5.25. Effect of wind on an isolated tree
(photo by Leclerc).

acters), following human activity, that was potentially harmful to humans. Gradually the idea expanded to include anything harmful to the equilibrium and preservation of ecosystems.

If we refer to the "natural state" when studying natural environments, an average situation must be observed and recorded that is considered normal, or at least what are considered normal ranges of variation of different factors. But catastrophic situations can always occur, major and rare conditions such as those that lead to serious changes of the geological ages, for example, the transition from the secondary era to the tertiary era.

Catastrophes have more limited effects. Particularly violent volcanic eruptions have short-term, geographically localized effects, but they may also have a long-lasting impact on the vegetation, as when large quantities of CO_2 were emitted in some areas of Italy. Volcanic ash sometimes reaches the stratosphere and influences the terrestrial climate for about one or two years.

Other volcanic emissions, such as SO_2, H_2S, HCl, H_2SO_4, NOx, or sulphates, in the gaseous or aerosol state, have always exerted marked effects, but these substances are produced in overall much larger quantities by human activity.

4.3.8.2. Atmospheric pollution

- **Problems of classification**

It does not seem advisable to separate at the outset those molecules that are entirely of human origin. A certain difficulty arises from the physical state of pollutants, either gaseous, liquid, or solid, the last two being aerosols.

A product may in fact be more dangerous in the liquid state (often in an aqueous solution) than in the solid or gaseous state. Aerosols are present in particles or drops of a diameter of about one micron and are thus, in many atmospheric situations, practically as mobile as in the gaseous state.

One difficulty with chemical or physicochemical measurements arises from the existence of complex mixtures and coexistence in several states. There is therefore an advantage in using biological measurements.

- **Bioindication**

Given the complexity of the problem, and the existence of synergy, given that pollution must be understood overall, we can at most evaluate the whole according to measurable effects on living organisms (bioindicators) that are more or less sensitive and on the basis of tests.

In the case of atmospheric pollutants, the higher plants have means to defend themselves against sudden pollution: stomatal regulation, cuticular protection of leaves, and protection of branches by suber formation.

For a long time it has been noted that lichens are highly sensitive to atmospheric pollution (large contact area, absence of stomata and a proper cuticle) and can be excellent bioindicators. One difficulty, however, is recognition of species because of the great phenotypic variability of lichens (Table 5.19).

Despite their natural protections, certain higher plants are still particularly sensitive (Table 5.20), but others are highly resistant (Table 5.21). Trees have especially been studied because of their economic importance and their sensitivity to winter pollution (Table 5.22).

- **Physiological effects**

Many studies have attempted to explain the effects of atmospheric pollutants on growth and various parameters of functioning of plants. The

Table 5.19. Average concentration of pollutant SO_2 in lichen species (Deruelle, 1979; Kershaw, 1985; Arndt et al., 1987, based on their results and those of many other authors)

Mean SO_2 conc., $\mu g/m^3$	Epiphytic lichens		Epilithic lichens	
	Eutrophic bark	Non-eutrophic bark	Alkaline substrate	Acid substrate
> 125	*Lecanora conizaeoides* *L. expellans*	*L. conizaeoides* *Lepraria incana*	alliance of *Lecanorion dispersae*	alliance of *Conizaeoidion*
70	*Buella canescens* *Physcia adscendons*	*Hypogymnia physodes* *Lecidea scalaris*		alliance of *Conizaeoidion* *Acarospora fuscata*
60	*Buella canescens* *Xanthoria parietina* *Physcia arbicularis* *Ramalina farinacea*	*Hypogymnia physodes* *Evernia prunastri*	alliance of *Xanthorion*	
50	*Pertusaria albesans* *Physconia pulverilenta* *Xanthoria polycarpa* *Lecania cystella*	*Parmelia caperata* *Graphis elegans* *Pseudoevernia furfuracea*		
40	*Physcia aipolia* *Ramalina fastigiata* *Candelaria concolor*	*Parmelia caperata* *Usnea subfloridana* *Pertusaria hemispherica*	alliance of *Xanthorion* more rich in species	several *Cladonia*
< 30	*Ramalina calicaris* *Caloplaca aurantiaca*	*Lobaria pulmonaria* *Usnea florida* *Teloschistes flavicans*	up to 20 species of *Xanthoria*	great diversity but disappearance of *Lecanora conizaeoides*

The species indicated are present or appear below the values indicated.

Table 5.20. Some terrestrial phanerogams that can serve as pollution controls by their sensitivity (results from many authors, Steubing, 1976; Ernst and Joosse-van-Damme, 1983); Arndt et al., 1987; Rabe, 1990; Schulze and Stik, 1990; Schubert, 1991; in Larcher, 1995)

Pollutant	Species	Particularly sensitive variety
SO_2	*Populus tremula* (European aspen)	
	Medicago sativa (lucerne)	
H_2O	*Pseudotsuga menziesii* (douglas fir)	
	Spinacia oleracea (spinach)	
HF, F2	*Prunus armeniaca*	
	Gladiolus communis (gladiolus)	Snow Princess, Shirley Temple
HCl	*Syringa vulgaris* (lilac)	
	Fragaria vesca (strawberry)	
NH_3	*Taxus baccata* (yew)	
	Brassica oleracea (cabbage)	Choux-fleur Le Cerf
NOx	*Apium graveolens* (celery)	
	Petunia × hybrida	
O_3	*Nicotiana tabacum*	Bel W3
	Phaseolus vulgaris (bean)	Sanilac, Pinto III, Tempo
PAN	*Petunia × hybrida*	
	Phaseolus vulgaris (bean)	
	Poa annua (annual bluegrass)	Provider, Astro
Ethylene	*Petunia × hybrida*	White Joy

Table 5.21. Trees and bushes tolerating severe urban SO_2 pollution (Krusmann, 1970; Dassler, 1991, in Larcher, 1995)

Highly tolerant	Moderately tolerant
Broad leaves	
Acer platanoides (Norway maple)	*Castanea sativa* (chestnut)
Buxus sempervirens (boxwood)	*Ginkgo biloba*
Celtis australis (southern nettle tree)	*Magnolia hypoleuca*
Gleditschia triancanthos	*Populus candicans*
Platanus × hybrida (plane tree)	*Robinia pseudoacacia* (locust tree)
Quercus (several species of oak)	
Sophora japonica	
Conifers	
Juniperus (several species of junipers)	*Chamaecyparis* (several species)
Picea omorika (Serbian spruce)	*Picea pungens* f. *glauca* (blue spruce)
Taxus baccata (yew)	*Pinus mugo*
Bushes	
Calluna vulgaris (common heather)	*Thuja* (several species of thuja)
Erica carnea (spring heath)	*Berberis* (several red currants)
Gaulthieria shallon	*Forsythia intermedia*
Ligustrum (several species of privets)	*Prunus laurocerasus* (cherry laurel)
Sambucus nigra (black elder)	*Weigelia florida*

Table 5.22. Parameters that can explain changes in wood, under pollution stress
(Liese et al., 1975; Keller, 1980; Halbwachs and Wimmer, 1987; Wimmer and
Halbwachs, 1992, in Larcher, 1995)

Parameter	Conifer wood	Wood of deciduous trees
Cambial growth	−	−
Proportion of sclerenchyma	+	−
Thickness and density of walls	−	±
Length of fibres		−
Tracheids or vessels per unit of area	+	+
Diameter of tracheids or vessels	−	−
Number of bordered pits of tracheids	+	
Diameter of bordered pits of tracheids	−	
Number of rays per unit of area	+	+
Number of resin canals	+	
Reserve substances	x	
Proportion of cellulose	x	x
Proportion of lignin	−	x
Other parietal substances	+	x

+, increase. −, decrease. x, no apparent change.

effects of high CO_2 levels on trees have been the focus of research in many laboratories, because of the significance of photosynthesis for physiology and the essential impact that growth in biomass of trees has in a strategy of control against increase in atmospheric CO_2. The effects of CO_2 are visible at several levels with respect to an individual as well as a community (Table 5.23). Their interaction with hydric equilibrium is also very important (Fig. 5.19).

Of course, it is essential to take into account all the pollutants and their effects on the different morphological, histological, and physiological parameters of trees (Table 5.22), cultivated plants, and all plants (Table 5.24), using global models. Certain traits of plant physiology are particularly sensitive to pollution and can serve as controls in the study of its effects (Table 5.25).

Atmospheric pollutants are found, either deposited or diffused, in the water and soil, where they eventually undergo chemical modifications that are not necessarily beneficial.

4.3.8.3. Pollutants of water and soil
Pollutants of concern are not only general pollutants of the atmosphere, but also products of phytosanitary treatment, such as herbicides and pesticides, part of which remain for some time in the atmosphere (in case of sprays) but are later found at much lower strata.

The subject may also be extended to the impact of pollutants on buildings and monuments that are sensitive not only to acid rain and aerosols, but also to certain agricultural treatments. Sprays and scattering of nitrogenous fertilizers on intensive cultivations in Champagne led to

Table 5.23. Possible responses to doubling of CO_2 level (plants and communities)
(data from many authors, in Larcher, 1995)

Morphogenesis and development	
Thickness of mesophile	0 or +
Leaf surface/dry matter	0 or −
Leaf area index	+ or x
Stem growth	+
Density of ramification	+ or −
Root growth	+
Root/stem ratio	+ or −
Abundance of flowers	+ or −
Seed size	+
Life span	+ or x
Energy and hydro-mineral regulation	
Net photosynthesis	+, and C3 > C4
Rubisco	0 or −
CAM	+
Dark respiration	+ or −
Sugar transport	+ or x
Dry matter production	+
Transpiration	−, and C4 > C3
Water use efficiency	+
Nitrogen use efficiency	0
Level of minerals in dry matter	−
Production of dry matter in conditions of heat, salinity, or drought	+
Biocoenosis	
Use of light	+ or −
Use of water	+ or −
Use of minerals	+ or −
Competitiveness of whole plant	+ or 0 (C3 > C4)

+, increase. −, decrease. x, uncertain. 0, no change.

the growth of a vast layer of nitrophilous lichens on the church of Notre Dame de l'Epine in the 1980s.

When collecting data, we must also take into account that certain pollutants may be carried very far by wind (Table 3.8). Aerosols may be found more than 1000 km from their places of industrial emission.

In the case of soils, heavy metals and pesticides are particularly important. Heavy metals are most of the time trapped in the clay-humus soil complexes. Pesticides, however, undergo some biotransformation and percolation, which brings them, in a more or less modified form, to running water, which is consumed by animals and humans. The study of water pollution is therefore essential.

Some bioindicators are quite specific to certain types of organic or mineral pollution in the case of aquatic angiosperms (Table 5.24). Herbicides and pesticides have characteristic impacts on biodiversity and algal successions (Fig. 5.26) (Giacolou and Echaubard, 1987).

Table 5.24. Aquatic angiosperms serving as pollution controls
(Nobel et al., 1983, in Larcher, 1995)

Pollutant	Species	Sensitivity threshold	LD_{50} (ppm)
Phenol	*Potamogeton lucens*	0.2	0.6
	P. coloratus	0.6	
	P. crispus	0.6	
O-cresol	*P. lucens*	0.2	
	P. coloratus	0.6	0.7
	P. crispus	1.1	1.1
KH_2PO_4	*P. alpinus*	0.2	2.0
	Helodea canadensis	0.5	< 50
NH_4Cl	*P. coloratus*	< 5	15
	P. crispus	< 5	
	Ranunculus fluitans	25	
H_3BO_3	*E. canadensis*	< 1.0	10
	Myriophyllus alternifolium	< 2.0	5
	R. penicillatus	< 1.0	10
Lead	*P. crispus*	2.1	
	H. canadensis	10.4	
	P. lucens	10.4	
Cadmium	*E. canadensis*	0.01	0.6
	P. crispus	0.01	0.6
	P. lucens	0.01	0.6
Copper	*P. crispus*	< 0.03	0.06
	E. canadensis	0.006	0.3
	P. lucens	0.06	> 0.3
Zinc	*E. canadensis*	< 0.07	3.3
	P. lucens	0.7	6.5
	P. crispus	4.9	6.5

Various phenol derivatives may be biodegraded into products that are equally or even more dangerous, which are then found in lakes or rivers. Many of these pollutants can in theory be quite easily measured by physiochemical procedures (e.g., HPLC), but it may be useful to detect them by studying the growth of *Scenedesmus quadricauda* in polluted water.

Present research trends aim to develop biocaptors for quick and sensitive detection of certain herbicides from photosynthesis, or of volatile organic compounds used in industry, or even of heavy metals (Naessens, 1998). Heavy metals are extensively studied in ecology and physiology because of their toxic effects, and because they are difficult to eliminate from soil and water.

4.3.8.4. Heavy metals

Heavy metals must first be detected and measured. This is a difficult problem in soil, given the multiplicity of forms they may take, particularly divalent, trivalent, and higher forms, but it is often easier in water. In an aquatic environment, many species of mosses accumulate heavy metals

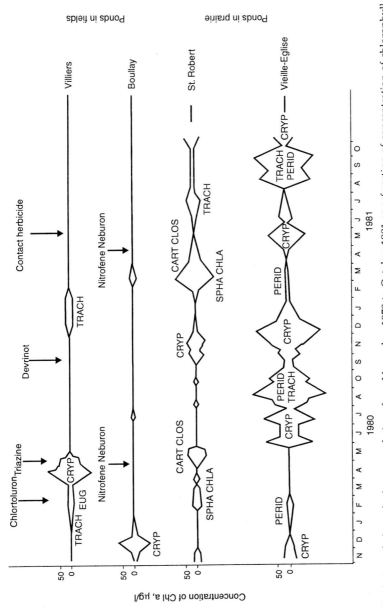

Fig. 5.26. Evolution of phytoplankton populations from November 1979 to October 1981 as a function of concentration of chlorophyll pigments in four ponds
(Giacolou and Echaubard, 1987).

TRACH = *Trachelomonas*, EUG = *Euglena*, CRYP = *Cryptomonas*, SPHA = *Sphaerellopsis*, CHLA = *Chlamydomonas*, CART = *Carteria*, CLOS = *Closterium*, PERID = *Peridinium*.

Table 5.25. Biochemical and physiological characters showing stress due to atmospheric pollution (Hartel, 1976; Horsman and Wellburn, 1976; Jager, 1982; Darral and Jager, 1984; Weigel et al., 1989; Grill et al., 1989; Cape and Vogt, 1991, in Larcher, 1995)

Indicator	Pollutant	Change
Enzymes		
Peroxidase	F, HF, SO_2	+
Polyphenol oxidase	SO_2, NO_2, hydrocarbons	+
Glutamate dehydrogenase	SO_2, NOx	+
Rubisco	SO_2	−
Nitrite reductase	SO_2, NOx	−
Superoxide dismutase	acid rain, O_3	+
Stress metabolites		
Ascorbic acid	non-specific	+
Glutathion	SO_2	+
Polyamine	non-specific	+
Ethylene	non-specific	+
Metabolism		
Energy charge	non-specific	−
Photosynthesis	non-specific	−
Optic foliar reflectance	O_3, SO_3, acid precipitation	−
Turbidity test*	acid precipitation	+

+, increase. −, decrease.
*Turbidity test: turbidity of hot water that has circulated through conifer needles.

Table 5.26. Mosses indicating heavy metals by capacity for accumulation (Arndt et al., 1987; Tyler, 1990)

Element	Species
Lead	*Bryum pseudotriquetrum*
	Dicranella varia
	Mielichhoferia nitida
	Philonotis fontana
	Fontinalis squamosum
	Scapania undulata
Copper	*Calypogeia muelleriana*
	Merceya ligulata
	Mielichhoferia elongata
	Mielichhoferia nitida
	Cephalozia bicuspidata
	Oligotrechum hercynium
	Pohlia nutans
Zinc	*Cephalozia bicuspidata*
	Pohlia nutans
	Bryum pseudotriquetrum
	Dicranella varia
	Mielichhoferia nitida
Nickel	*Oligotrechum hercynium*

and can serve as bioindicators of cumulative pollution (Table 5.26). In a terrestrial environment, metallophytes, such as *Armeria maritima* and *Viola calaminaria*, allow accumulation of heavy metals and serve to characterize soils (Table 5.27).

The presence of heavy metals is widespread in certain countries, particularly in Eastern Europe. It is a matter of concern, of course, near mines and mineral treatment zones. Heavy metals are found more particularly in certain types of soils (Fig. 5.27), in industrial zones, and near waste incineration plants (Fig. 5.28).

The effect of stress is characteristic in some cases. For example, during the action of soluble aluminium on a sensitive variety of wheat, there is rapid ulceration of young roots, visible under scanning electron microscope, then the aluminium penetrates the meristem and blocks root growth (De Lima and Copeland, 1994).

In contact with heavy metals, the forms of which solubilize especially in acid soils, plants develop strategies of adaptation and/or defence.

Table 5.27. Some metallophytes with high level of heavy metals or comparable elements (mg/kg dry matter) (data from many authors: Duvigneaud and Denaeyer-De Smet, 1973; Ernst, 1976, 1990; Baumeister and Ernst, 1978; Steubing et al., 1989)

Species	Distribution	Element	Conc.	Accumulation factor*
Eichhornia crassipes (water hyacinth)	Tropical waters	Fe	14,400	10
Minuartia verna	Central Europe			
leaves		Cu	1,030	147
roots			1,850	109
leaves		Pb	11,400	950
roots			26,300	970
leaves		Cd	348	3,480
roots			382	3,820
Thlaspi coerulescens	British Isles			
leaves		Zn	25,000	208
roots			11,300	140
Jasione montana				
leaves	British Isles	As	31,000	
Mechovia grandiflora				
leaves	Congo Basin	Mn	7,000	7
Acrocephalus robertri				
leaves	Congo Basin	Co	1,490	50
Psychotria douarrei	New Caledonia			
leaves		Ni	45,000	
roots			92,000	
Pearsonia metallifera	East Africa	Cr	490	98
			1,620	162
Astragalus racemosus leaves	North America	Se	15,000	

*According to Duvigneaud (1967): level in polluted environment/level in normal environment.

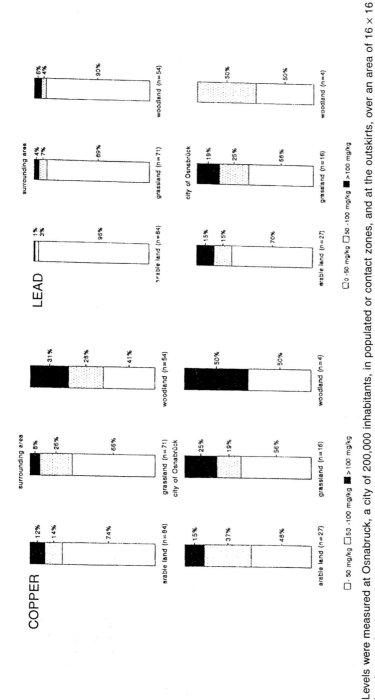

Fig. 5.27. Distribution of metals in some soils of a town in northwestern Germany and neighbouring areas
(Bloemen et al., 1995).

Levels were measured at Osnabruck, a city of 200,000 inhabitants, in populated or contact zones, and at the outskirts, over an area of 16 × 16 km, in grassland, arable land, and woodland. Woodland soils were particularly acidic (average pH 3.9, sometimes below 3). Lead pollution was greater in the city but more intense in the outskirts, especially in the forest, probably in connection with the acidity. Copper pollution declined rapidly towards the outskirts but was occasionally high.

The problem occurs for copper, zinc, lead, mercury, cadmium, and nickel in industrial zones, and for aluminium the problem is more extensive: aluminium is the most abundant metal on earth and is found in most acid soils.

The plant may try to prevent the entry of lead and other heavy metals. If this is not possible, the plant attempts to confine the heavy metal to the roots rather than allowing it to reach the stems or leaves (Figs. 5.28, 5.29). If the metals reach the leaves, the plant sets up a system of excretion, at the trichomes, for example, or distributes the pollutant in the tissues or cells (Fig. 5.30). At the molecular level, plants may "neutralize" heavy metal ions using metallothioneins.

The physiological effects of heavy metals occur mostly in plant development and photosynthesis. Aluminium, for example, accumulates in the nuclei of root meristem in wheat (Galsomies et al., 1992) and blocks root growth. Because of this, it causes high drought-sensitivity in crops.

Heavy metals have an effect at several levels, particularly aluminium and nickel, which are both useless to living things, even at low doses, unlike copper, for example (Table 5.28, Fig. 5.31). Many of the effects are indirect and compete with, or inhibit, the absorption of useful elements such as potassium and calcium. In combination with these effects, certain elements may be directly harmful (Fig. 5.32).

The example of aluminium-boron interaction shows that two or more metals together have synergistic as well as antagonistic effects. The interactions involve not only stress from metals but also all other types of stress.

4.3.9. Combined effects

Combined effects of stresses have been studied systematically and can be grouped into types. The interactions of drought and high CO_2 were studied in many laboratories. Water stress is reduced if CO_2 is abundant, 700 vpm for example, and the water potential of soils (Schwanz et al., 1996) for plants thus treated will decline less than for potted plants in a normal atmosphere.

In English oak (*Quercus robur*) (Table 5.29), the plants adapted to CO_2 and drought have a fairly low potential at the end of the night, which shows the beginning of a xerophytic character. In these plants an increase is observed in carotenoids, which protect against photooxidation. In the same conditions (Table 5.30), the biomass of young pines is better.

At higher CO_2 levels (double, for example), the better diffusion of CO_2 towards substomatal chambers allows a relative closure of stomata, and thus the plant loses less water under drought stress.

During a drought-ozone interaction in soybean (Table 5.31), ozone is observed to mitigate the leaf damage caused by water stress, at least when the plant is young. However, the water stress cannot prevent the precocious ageing caused by ozone.

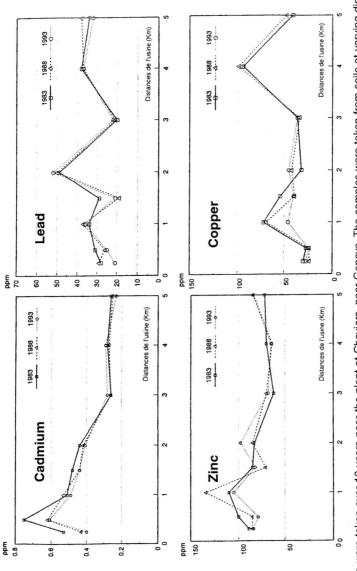

Measurements were taken over 10 years near the plant at Cheviers, near Geneva. The samples were taken from soils at varying distances from the plant and exposed to winds from the west (dominant). Cadmium showed the best curve of decrease with distance. Here it can be considered an indicator of soil pollution by heavy metal. Zinc, which has a physicochemical behaviour similar to that of cadmium, presents a profile that is not clear. The curves of copper and lead reflect other sources of pollution.

Fig. 5.28. Medium-term evolution of heavy metal pollution near an incineration plant
(*Robin et al., 1995*).

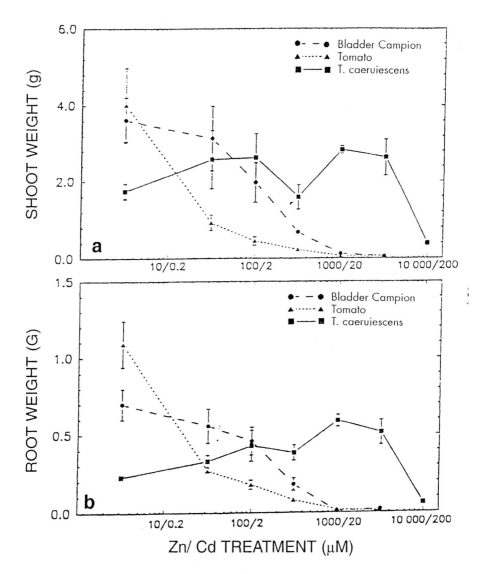

Thlaspi caerulescens (J and C. Presl.) is a cruciferous plant that hyper-accumulates cadmium and zinc. At high pollution levels, root biomass was still large, which must be an adaptive character. The plants came from Prayon (Belgium). Silene vulgaris (Moench) Gorcke L., bladder campion, is a zinc-tolerant ecotype collected at Palmerton (Pennsylvania, USA). It tolerated medium levels of pollution well but could not live in extremely polluted conditions. Lycopersicon lycopersicum (L.) Karsten, tomato, is intolerant of zinc and cadmium. It was very quickly affected even at low pollution levels. Cadmium and zinc were always applied in the same ratio, 1/50, close to values observed at Palmerton.

Fig. 5.29. Effect of cadmium and zinc pollution on roots and shoots of three plants
(*Brown et al., 1995*).

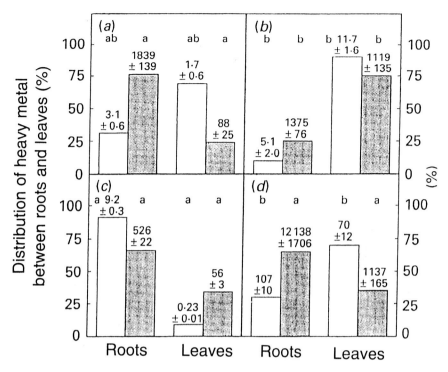

The heavy metal content of all the leaves and roots is 100%. The proportions are represented by bars (white for control, grey for treated parts). The concentrations (nmol/g fresh matter) are indicated in the columns with the standard deviations evaluated over 6 to 10 measurements. Different letters indicate the significant differences at P < 0.01. (a) cadmium; (b) molybdenum; (c) nickel; (d) zinc. In case of stress, the roots are observed to retain a large part of the pollutants, except in the case of molybdenum, which may be accumulated in high concentrations in leaves as well as roots, but the compartment is larger. Nickel, which has no nutritive value for plants, is blocked at the roots even in the control.

Fig. 5.30. Distribution of heavy metals in barley plants between roots and leaves
(Brune et al., 1995).

Very often, there is a negative interaction between drought stress and cold stress. A relative dryness allows the plant to better resist cold and the reverse is also true. For example (Table 5.13), an acclimatization to drought reduces the effects of cold in maize. In particular, there is better recovery of photosynthetic activity.

The effects of cold and effects of intense light are commonly studied together. For example, in cucumber, an effect of photoinhibition is observed in the presence of cold, affecting mainly the quantum yield of photosystem I (Fig. 5.24).

The great majority of studies on stresses and their interactions have focused on cultivated plants and thus on the terrestrial environments, but we must not ignore what happens in an aquatic environment.

Table 5.28. Photosynthesis and aluminium (Moustakas et al., 1996)

Parameter	Treatment 1	Treatment 2	% change
Net CO_2 assimilation (mmol $CO_2/m^2/s$)	22.17 ± 0.20	16.37 ± 0.24	–26
Transpiration (mmol $H_2O/m^2/s$)	4.22 ± 0.004	3.04 ± 0.13	–28
Intercellular CO_2 conc. (ml CO_2/l)	282 ± 1.10	253 ± 1.60	–10
Stomatal conductance (mmol $H_2O/m^2/s$)	420 ± 2.20	244 ± 3.10	–42
Quantum yield of CO_2 fixation	0.027 ± 0.00	0.0195 ± 0.00	–28

Measurement of leaf gas exchanges in controls (OmM) and in *Thinopyrum bessarabicum* subjected to stress for 48 h by 1 mM Al^{+++} at pH 9 (from KAl(SO$_4$)$_2$). *Thinopyrum* is a Mediterranean halophyte. The pH used is not far from that of the natural environment and prevents the interactive effect of acidity.

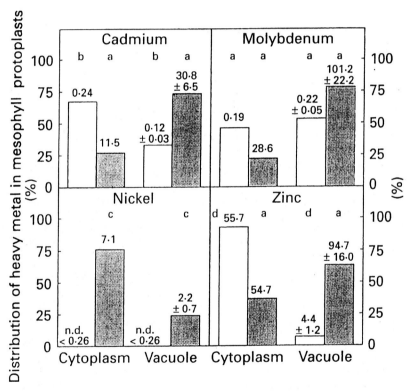

The effects of accumulation were studied on barley, protoplasts of which were prepared after culture in a "normal" medium or in one with heavy metals (100 μmol/l for Cd or Ni; 400 μmol/l for Mo or Zn). The enzyme α-mannosidase serves as a vacuolar marker. The figures on the columns indicate the concentrations in nmol/g fresh matter. White columns represent the control. Different letters indicate different distributions of heavy metal, at P < 0.01.

Fig. 5.31. Distribution of heavy metals in mesophyll cells of the leaf
(*Brune et al., 1995*).

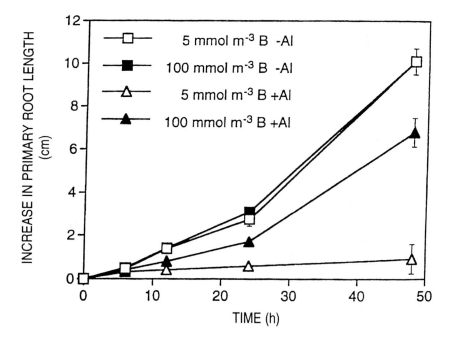

One of the classic effects of aluminium stress is inhibition of root growth, studied here in squash (*Cucurbita pepo* L). It was previously determined that in the absence of Al, doses of boron compatible with optimal root growth ranged from 5 mmol to 100 mmol/m³. A dose of 1.5 mol/m³ AlCl₃ and 6H₂O (which corresponds to 44 mmol/m³ of free aluminium) was added to the nutrient medium. Boron treatment led to restoration of two-thirds of root growth.

Fig. 5.32. Antagonism between the effect of aluminium stress and boron treatment
(Lenoble et al., 1996).

Table 5.29. Combined effect of CO₂ increase and summer water stress on leaves of young English oak (*Quercus robur* L) (Schwanz et al., 1996)

Parameters	A-W	A-D	E-W	E-D
Water content (%)	58.6 ± 1.18 a	6.10 ± 3.7 a	61.6 ± 5.4 a	59.6 ±2.4 a
Water potential at end of night (MPa)	–0.64 ± 0.15 a	–0.79 ± 0.24 ab	–0.69 ± 0.20 a	–0.91 ±0.26 b
Leaf mass/surface ratio (g/m²)	84 ± 8.7 b	74.7 ± 5.3 a	74.7 ± 12.8 a	68.5 ±5.7 a
Chlorophyll (mg/g)	3.39 ± 0.32 a	3.45 ± 0.96 a	4.17 ± 0.84 ab	4.96 ±0.93 a
Carotenoids (mg/g)	0.75 ± 0.09 a	0.81 ± 0.21 a	0.87 ± 0.15 ab	0.93 ±0.19 b
Soluble proteins (mg/g)	101.5 ± 15.9 a	89.9 ± 23.1 a	93.3 ± 21.5 a	99.5 ±20.2 a

Oaks placed for 5 months in the greenhouse at 700 µl CO₂/l (E) or ambient air (A) and subjected (D) or not subjected (W) to hydric stress from 13 July to 26 August 1993. In each range, different letters indicate a significant difference (P ≤ 0.05).

Table 5.30. Combined effects of CO_2 increase and summer water stress on needles of young pines (*Pinus pinaster*) (Schwanz et al., 1996)

Parameters	A-W	A-D	E-W	E-D
Water content (%)	69.9 ± 4.5 b	69.7 ± 3.4 b	66.7 ± 5 ab	63.6 ± 6.6 a
Water potential at end of night (MPa)	−0.56 ± 0.41 a	−2.59 ± 0.83 b	−0.83 ± 0.23 a	−2.40 ± 0.23 b
Leaf mass/surface ratio (g/m²)	236 ± 10 a	226 ± 10 a	215 ± 10 a	252 ± 8 a
Chlorophyll (mg/g)	1.84 ± 0.22 a	1.99 ± 0.54 a	1.57 ± 0.67 a	1.42 ± 0.25 a
Carotenoids (mg/g)	0.48 ± 0.02 b	0.46 ± 0.07 ab	0.39 ± 0.07 a	0.39 ± 0.10 a
Soluble proteins (mg/g)	10.0 ± 2.4 c	8.34 ± 1.64 bc	6.06 ± 3.14 a	6.45 ± 1.891 ab
Total weight (g/plant)	25.7 ± 2.9 a	27.4 ± 3.1 a	36.4 ± 6.2 b	38.7 ± 2.8 b

Pines placed for 6 months in a greenhouse at 700 ml CO_2/l (E) or ambient air (A) and subjected (D) or not subjected (W) to hydric stress from 16 June to 26 August 1994. In each range, different letters indicate a significant difference (P ≤ 0.05).

Table 5.31. Example of combined effects of drought and ozone stress (Vozzo et al., 1995)

DAP	CF WW	CF WD	1.5 WW	1.5 ND	2.1 WW	2.1 WD
53	0	0	2	0	12	2
61	3	2	12	7	32	18
67	0	0	9	1	24	6
73	2	0	4	1	22	8
82	2	0	20	2	31	21
88	4	0	33	6	47	32
95	4	0	38	8	56	44
102	0	0	44	20	60	59
111	2	0	57	39	74	62
117	32	21	85	89	91	89
125	18	22	88	100	100	100
130	24	35	98	97	100	100
138	81	89	100	100	100	100
140	100	100	100	100	100	100

The study was conducted in young soybean plants in the field, on the median leaflet of the leaf of the 6th node. The ozone treatment began on the 27th day, and the hydric treatments from the beginning. DAP = days after planting. CF = air filtered through activated carbon (no ozone); 1.5 = normal ozone in the air × 1.5. 2.1 = normal ozone in the air × 2.1. WW = well-watered. WD = only part of the natural precipitation. Statistical results were recorded on 6 plants per treatment. Significant effects (P = 0.05) were observed on all the dates.

Effects of photoinhibition, and thus intense light stress, are well known in freshwater as well as marine algae. From the point of view of interactions, the joint effects of pH and light are particularly significant in fresh water. Much of fresh water is more or less acidic and thus has little or no bicarbonate. The only CO_2 reserve is dissolved CO_2. In intense light, plants may thus quickly find themselves in a situation of CO_2 deficiency, which leads to photoinhibition due to the absence of sinks for electrons emitted by the photosystems. The same situation is observed in a marine alga, for which life in more or less acidic water represents a typical stress. In the case of the tropical red alga *Eucheuma*, the appearance of oxygenated water is observed, which indicates photoinhibition (Fig. 5.33).

A stress often does not come alone, but we must also remember that combined effects involve abiotic stresses and biotic stresses. We must thus analyse stress caused by other living things, such as animals and pathogens, as well as what is increasingly interesting to study, allelopathy, i.e., the harmful effect that plants have on other plants.

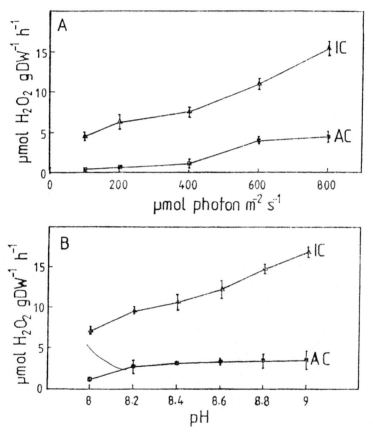

The study was conducted on a large, highly ramified Rhodophyceae found in warm seas: *Eucheuma denticulatum*, growing before the experiment at a daily growth rate of 3.5 to 4.7% under 350 μmol photons/m²/s. (A) Algae in pH 8.2, effect of light. (B) Algae under 400 μmol photons/m²/s, effect of pH. The production of H_2O_2 was measured in the presence (IC) or in the absence (AC) of 0.3 mM of sodium azide. The azide was used to inhibit catalase and haemic peroxidase in order to estimate the maximum production of H_2O_2. The means and standard deviations were calculated over four parallel measurements.

Fig. 5.33. Production of hydrogen peroxide by an alga subjected to pH and intense light stress
(*Mtolera et al., 1995*).

5. STRESS CAUSED BY OTHER LIVING THINGS

In this section, we analyse the action of animals and disease-causing organisms, such as fungi, bacteria, and viruses. Emphasis is also given to stresses due to other plants, called allelopathy.

5.1. Introduction

From early times, cattle herders observed that some plants were not grazed by animals. Similarly, gardeners are aware that some plants are unfavourable to others, and also that resistance to insects depends on the species and varieties of plants. There are in effect interactions at several levels. They may be simple, between plant and plant or between plant and herbivore, or they may be three-way or more complicated interactions.

Some studies integrate the primarily observational earlier studies and experimental studies using modern methods of analysis. Examples are the research of the Ecole Nationale de Grignon (Guyot et al., 1955) or observations on the effect of mugwort through the intermediary of soil or even air (Muller, 1965; Fig. 5.34).

Fig. 5.34. Allelopathic effect of mugwort plants in the American Northwest
(Muller, 1965).

Various cases of interaction are briefly presented in this section before we examine the extent of stress caused by other living things and explanations of the physiology of stressful interactions.

Complex interactions between plants involve more than two species and comprise allelopathy and a combination of biotic and abiotic factors (Shevtsova et al., 1995). Plants interact with animals, fungi, and bacteria. Fungi can, independently of parasitic effects, create active substances against herbivores (Sterner, 1995). On the other hand, there are now plant substances known to be antifungal or antibiotic (Pare et al., 1993; Moujir et al., 1993).

Plant-herbivore relationships vary according to the type of defence, physical or chemical, and involve all the intermediate effects between toxicity (Katz, 1990) and simple repulsion (Ndungu et al., 1995). Also, there are plant-herbivore-carnivore or similar relations (Heil et al., 1997) or plant-plant-herbivore-carnivore relations (Hacker and Bertness, 1996). Plant-herbivore interactions give rise to phenomena of evolution (Weintraub and Scoble, 1995).

Plant-plant interactions, apart from manifestations of allelopathy and competition for light, also include host-parasite relationships with a varying number of partners (Pennings and Callaway, 1996). In addition to classic symbiosis, we must also consider other positive plant-plant interactions (Hacker and Bertness, 1996).

Defence (or attack) has a cost. The plant must expend energy to synthesize complex substances (Vrieling and Van Wijk, 1994) or even support an army of defenders (Heil et al., 1997) or facilitate the action of an "involuntary defender" insect (Eigenbrode et al., 1995).

5.2. Plant-Plant interaction

Plant-plant interaction is based essentially on **allelopathy**. The subject is discussed as a whole in a publication by the American Chemical Society (Indergit et al., 1997), but we must also take into account that, as with people, plants are not necessarily acting in a hostile manner. They may simply be competing for use of the light and other nutrients, including CO_2 (Reekie, 1996). They may even have positive interactions that are beneficial to at least one of the partners.

5.2.1. Allelopathy

Plant-plant interaction in which one plant attempts to get rid of another is observed in the terrestrial as well as aquatic environment (Vance and Franko, 1997). Apart from competition for nutrients, a plant may act either through the soil, which seems to happen most frequently, or through the air, by emitting volatile organic compounds, or through pollen (Murphy and Aarssen, 1995), in an action similar to that of an aerosol.

Allelopathy is perhaps particularly well developed in the tropics (Anaya et al., 1995), where competition is especially complex. It is often studied in the context of the action of ornamental or aromatic species against various edible vegetables or fruits, for example, the effect of French marigold on lettuce and radish (Kaul and Bedi, 1995). Moreover, many wild species are sensitive (Seidlova and Sarapetka, 1997) and sometimes secondary substances of a species have quite a wide spectrum, as with the lupanes, triterpenes of Fabaceae that can also act at very low doses (sometimes at 10^{-9} M) (Macias et al., 1994).

An allelopathic species can also act against germination of another allelopathic species (Van Staden and Grobbelaar, 1995). Allelopathy is a factor of biological evolution because of its selection pressure. A classic case is that of the walnut, which originated in the mountains of Central Asia. In its area of origin, the flora of the undergrowth are adapted to the walnut but, in Western Europe, the walnut eliminates the indigenous herbaceous species in its vicinity.

This phenomenon is observed in other species such as *Anthoxanthum odoratum*, a familiar plant in the prairies of Western Europe, which by means of allelopathy invades *Zoysia* grasslands in Japan (Yamamoto, 1995). The effect of an allelopathic substance can be as intense as that of a synthetic herbicide (Anaya et al., 1995).

5.2.2. *Interaction in the context of competition for nutrients*
Competition is particularly active for light. That is what the physiology of stress amounts to. A plant can detect the presence of neighbours by the quality of light they reflect on it and by the consequent modification of the ratio of red light to far red light (R/FR), and it aims ultimately to avoid the shade caused by neighbouring plants (Schmitt and Wulf, 1993).

There are similar **avoidance phenomena** at the soil in the context of root development. The roots of an individual absorb water and nutrients from the soil, and thus an impoverished area develops around it. Moreover, if the environment is heterogeneous (a mix of rich and poor patches of soil), two plants may compete to occupy a rich zone. The same applies to detection and exploitation of a water-rich area (Takahashi and Scott, 1993)

There is some specificity; for example, one plant may require a greater quantity of a given element. All of this is based on a system of information involving several types of signals (Aphalo and Ballare, 1995).

5.2.3. *Positive interactions*
The presence of a neighbouring plant B may protect or increase the well-being of a plant A. The protection may be mechanical, against access by herbivores. An example that covers these aspects is found in studies by Hacker and Bertness (1996). *Juncus gerardii* L facilitates the establishment of another plant of the salty marshes, *Iva frutescens* L., at a low level of the

strand, where it cannot survive on its own. This is an example of mechanical protection. When *Juncus gerardii* is grown experimentally, the photosynthesis of *Iva* plants diminishes and they are colonized by aphids (Fig. 5.35). Ultimately the plants stop growing and they and their aphids disappear. There is thus a hidden effect and another favourable effect. Another hidden effect occurs in relation to a carabid that is a natural predator on aphid (Fig. 5.36).

5.2.4. Interactions in parasite systems

Plants that are partial parasites can live entirely as autotrophs but develop better if they have established themselves on a host. Purely parasitic plants cannot live as autotrophs and must have a host.

In the case of partial parasitism, we can compare the impact of *Rhinanthus serotinus* and *Odontites rubra* on *Medicago sativa* (Matthies, 1995). The *Rhinanthus* is less effective as an autotroph than the *Odontites*. There are thus variable levels of partial parasitism. Lucerne attempts to defend itself in competition for light, and it can thus significantly slow down the growth of the partial parasite. The joint productivity of the host and parasite combined is less than the sum of the productivities of each.

An example of parasitism can be studied in the salty marshes of California (Pennings and Callaway, 1996). In these marshes, *Salicornia virginica* is the dominant species, and it is accompanied by *Limonium californicum* and *Frankenia salina* and other species. The parasite is *Cuscuta*

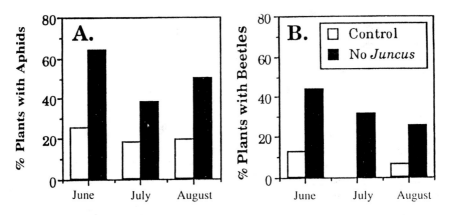

In a previous experiment it was shown, during measurements in June, July, and August, that photosynthesis of *Iva* (in μmol CO_2/m²/s) declines when the neighbouring *Juncus* plants are removed. In A we see that aphids become more numerous on *Iva* when the *Juncus* plants have been removed. In B, we see the same effect on the predator ladybirds. The *Juncus* plants exert a hidden effect.

Fig. 5.35. Positive interaction between *Juncus* and *Iva frutescens*
(*Hacker and Bertness, 1996*).

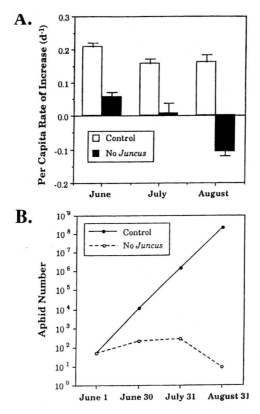

The study was conducted in cages that could be entered by aphids but not by ladybirds. The growth of the aphid population was slowed down and then the population declined when the *Juncus* plants were absent (A). (B) By mathematical integration and supposing there was no migration of aphids, the aphid population was observed to decline.

Fig. 5.36. Positive plant-plant interaction
(Hacker and Bertness, 1996).

saline, which prefers *S. virginica* to the other halophytes present. By manipulating the environment and observing intact parcels of the marsh, we can see that when there is severe infection by *Cuscuta* the plant community reacts by better development of *Limonium* and *Frankenia.* These two species are favoured in the competition with *Salicornia,* which is weakened. Nevertheless, the effect is not simple. In the lowest zones occupied by *Salicornia,* the *Limonium* and *Frankenia* are not favoured by parasitism on *Salicornia.* This can be interpreted either as a lower efficiency of the parasite due to the environment, or as a poorer adaptation of *Limonium* and *Frankenia* to an environment invaded more frequently by sea water. We must therefore place plant-plant interactions of different types in the context of interactions with the environment.

The interaction was observed in the Queyras at 1800 m altitude on land that was entirely bare during the previous year. The *Rhinanthus* has an erect inflorescence and is about 20 cm high.

Fig. 5.37. Interaction between a *Rhinanthus* plant and a member of the family Fabaceae *(photo by J.C. Leclerc).*

5.2.5. Interactions between plants and fungi or bacteria

Fungi themselves produce defence substances. For example, several Lactaria produce substances with an acrid taste to repel herbivores (Sterner, 1995). Other fungi produce antibacterial antibiotics or even toxins (e.g., triterpenes, oligopeptides) that are highly effective against mammals. It is important to remember, however, that fungi can have positive interactions with plants, chiefly in the form of mycorrhizal symbiosis.

Fungi also produce the equivalents of toxins against the plants they parasitize. However, plants react to the stress of the attack, sometimes in a hypersensitive way, by producing phytoalexins, many examples of which are now known in biochemistry and regulation (Fig. 5.38; Tsuji et al., 1992). Sometimes, the phytoalexins produced are in turn degraded by the attacking fungus (Bennett et al., 1994).

Other defences are essential, as with *Chenopodium ambrosioides*, which always produces, among its essential oils, terpenoids that have good antifungal properties (Pare et al., 1993). The same is true of antibacterial diterpenes produced by sage (Moujir et al., 1993).

Some situations are more complex, with only part of the defences being essential, such as plant defensins (antifungal or sometimes

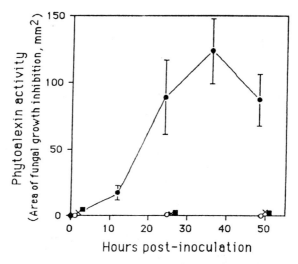

The quantity of phytoalexins produced by the plant is tested by the antifungal action of an extract on a culture of fungus.

Fig. 5.38. Induction of phytoalexin production
(Tsuji et al., 1992).

antibacterial peptides) (Broekaert et al., 1995). These defensins have structural analogies with those of insects and mammals (Fig. 5.39).

5.3. Plant-Animal Interactions

The interaction between plants and animals is fundamentally based on the plant-herbivore relationship. The herbivore has a problem of choosing the best plant for consumption. The plant attempts to avoid the herbivore by means of attack (toxins) or repulsion, or by diversionary means such as a lure, thus affecting the behaviour of the herbivore.

The herbivore may not always be inimical to the plant. Sometimes it disperses fruits or seeds (zoochory) or even, in the case of prairie grasses, favours the regrowth of the aerial parts, the meristems being at the base of the leaves. Often, herbivory plays a significant role in the relationships between plant and insect (or other animal) pollinator.

5.3.1. Degrees of defence
The defence of plants shows varying degrees and is accompanied by an animal adaptation that may include a defence of the herbivore against secondary consumers.

Many plants are toxic, for example many Ranunculaceae, among them species of *Aconitum* (Katz, 1990). Some aphids adapt themselves and absorb large quantities of aconitins and similar substances. Two species of *Aconitum* seem to be associated with two species of aphid, respectively, each adapted to (and accumulating) a certain type of toxin. Seasonal

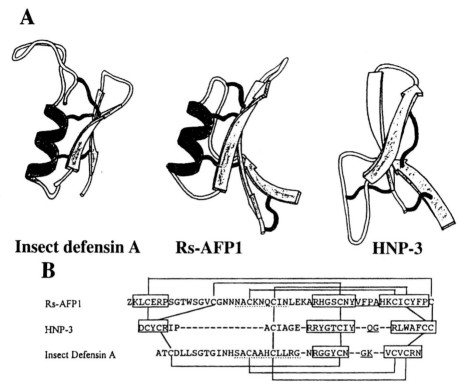

RS-AFP1 is a defensin of angiosperms, HNP-3 is a defensin of mammals. Defensins are small peptides with antimicrobial properties. They are widespread, relatively cheap to produce in terms of energy. They constitute a defence, at least partly inducible, that is perhaps more ancient than the evolutionary divergence between animals and plants. Defensins of plants are more antifungal. The sequences of amino acids are represented in the illustration, the boxed parts being the β strips. The underlined parts are the β bends, and the parts underlined with dotted lines are α helixes. The disulphur bridges are indicated. The a helixes may explain part of the effect, by ionophore properties (creation of canals) at the plasma membranes.

Fig. 5.39. A new type of antimicrobial peptide
(Broekaert et al., 1995).

variations may be observed, for example in the relationship between an Asteraceae and a deer (Hall et al., 1994). In this case herbivory occurs mostly in the winter and is linked to a weakening of volatile compounds (e.g., hexane, Fig. 5.40) and complex secondary substances. In summer, the plant must probably protect itself better in order to grow and reproduce.

In some plants the natural properties of repulsion probably have fairly wide spectra of action, so that plants that produce repellents vis-à-vis their herbivores can also be used to prepare effective extracts against ticks and weevils, as substitutes for synthetic chemical products (Ndungu et al., 1995).

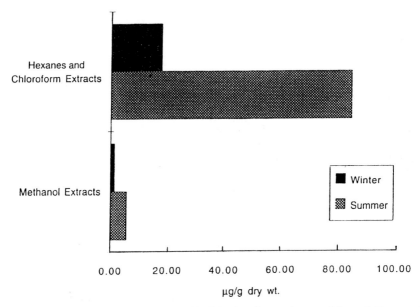

Herbivory occurs mostly in the winter and is linked to the weakness of the volatile organic compounds, particularly hexane and chloroform extracts. In summer, the plant must defend itself better to ensure survival of the species.

Fig. 5.40. Seasonal variations of volatile secondary metabolites
(Hall et al., 1994).

5.3.2. Plant-animal interactions in the aquatic environment

The same problems are found in aquatic and terrestrial environments. There are problems of selection, toxins, signals, repulsion, and digestibility. For example, a dinoflagellate exerts a specific predation on a unicellular red alga *Porphyridium* sp. If the cell wall of the alga is chemically modified (following nitrogen and sulphur deficiency in the algae), the predation is less severe and there is less enzymatic induction of glucosidases (Ucko et al., 1994). If the normal polysaccharides alone are applied after extraction, the enzymatic induction remains. The recognition thus seems to be based on the polysaccharides of the algal cell wall.

A particularly striking problem in the aquatic environment is that of epiphytism (Fig. 5.41). Many animal species live on large masses of algae, using them as a simple support, but when there is shade the composition of species changes, indicating that the "support" perhaps is not neutral.

5.3.3. Interactions between plant, herbivore and carnivore

In interactions between herbivore and carnivore, apart from fleeing from "enemies", herbivores also avoid enemies in their search for a new environment or a new plant. The plant benefits from the attack of

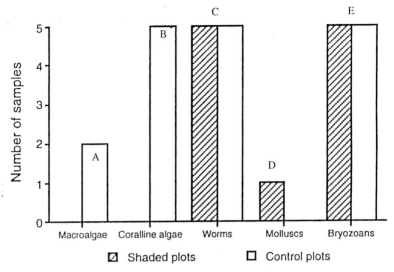

Fig. 5.41. Presence of epiphytic organisms on leaves of *Posidonia australis* in the shade and in control plots
(Fitzpatrick and Kirkman, 1995).

carnivores on herbivores. The secondary consumers are very often Coleoptera, ants, and parasitoid wasps. Different types of predators on herbivores are presented in the following examples.

Coleoptera constitute one category of predator. For example, cabbage is attacked by larvae of a nocturnal butterfly that is in turn attacked by several species of carnivorous Coleoptera (Eigenbrode et al., 1995). Some varieties of the cabbage have a glossy leaf coat instead of a normal wax cuticle with lipids of crystalline structure. In the greenhouse as well as in the field, caged cabbage with glossy leaves were attacked more than other cabbages by the three carnivores tested (Fig. 5.42). The explanation may be that the carnivores move more easily on the glossy leaves than on leaves with a normal wax cuticle surface. The plant may in this case facilitate attack by the carnivores.

Ants sometimes enter into a form of symbiosis with a tropical tree (Heil et al., 1997). The tree in question, *Macaranga tribola*, shelters ant colonies (Figs. 5.43, 5.44) by means of nutrient organs or food bodies derived from stipules. The trees inhabited by ants produce eight times as many food bodies as those without ants. Around 90% of the trees are inhabited by ants. This defence is costly for the tree in terms of proteins and lipids, but it is efficient, since the ants effectively discourage herbivores.

Parasitoid wasps have been extensively studied by scientists. The wasps often interact with plants and their choice of plant is determined by various factors. To the extent that there is a wasp-host specificity, and

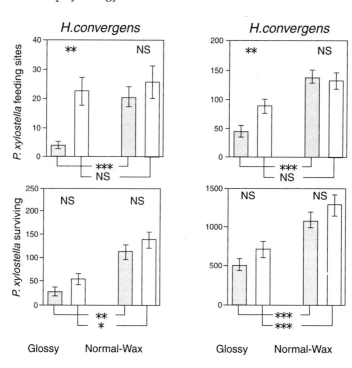

According to experiments in the greenhouse (at left) and in the field (at right), the predator *Hippodamia convergens* moves more easily and preys more effectively on glossy leaves, and the survival rate of the herbivorous larvae is lower.

Fig. 5.42. A relationship between plant, herbivore, and carnivore,
in which the plant uses a trick
(Eigenbrode et Coll, 1995).

a plant-herbivore specificity, there is a wasp-plant interaction in which the wasp recognizes the plant by means of an odour signal (Reed et al., 1995). In closed cages made of four chambers (two with plants, two with air), the parasitoid wasps preferred cages with leaves of cabbage and aphids. They showed the same preference when the leaves with aphids were removed and only the odour remained (Fig. 5.45). We can conclude from this that the parasitoids in the experiment were guided by the odour of the parasitized cabbage to look for aphids.

5.3.4. Problem of choice
The problem of choice exists with herbivores as well. For a long time this subject has been interesting in terms of, for example, cyanogenetic plants, certain typical species of which—trefoils and birdsfoot—present a bio-chemical polymorphism, interspecific and intraspecific. Can an insect (or

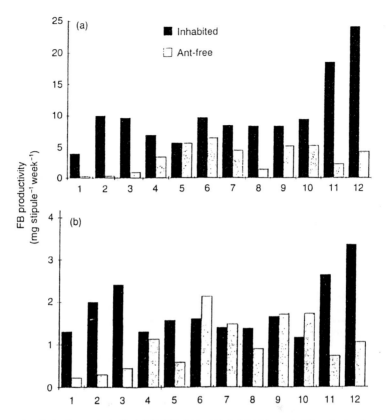

Twelve pairs of plants were studied, inhabited or not inhabited by ants. The plants of each pair were nearly of the same size. The same pairs were in (a) and (b), with the same numbers. (a) Total production of fruit bodies (FB) is given by all the pairs of stipules of a plant, in mg dry matter per week and per plant. (b) The productivity, i.e., the total production of a plant divided by the total number of stipules, in mg dry matter per week per stipule.

Fig. 5.43. Interaction between plant, herbivore, and protective ants
(*Heil et al., 1997*).

other type of herbivore) recognize and avoid a cyanogenetic plant? This choice is involved in the much larger context of food in relation to the needs of herbivores.

It is currently recognized that the large herbivores prefer dicotyledons to monocotyledons. This preference underlies the stability of a certain number of grassland systems. It also depends on the herbivore species, even if they are similar. For example, when given a choice of seven plants, *Cervus unicolor* prefers willow and *C. elaphus* prefers red trefoil (which is more digestible), and both avoid grasses (Semiadi et al., 1995). The two species have a ratio of similar digestibility, but *C. unicolor* eats a much larger quantity of plants rich in tannins, essentially willow, which is very

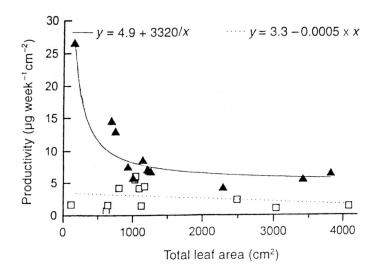

The productivity of fruit bodies (calculated in mg dry matter per cm^2 of leaf per week) was studied as a function of the total leaf area (cm^2) of plants. White squares represent plants without ants, and black triangles represent plants sheltering ants. The total leaf area gives an idea of the size of the plants. The regressions were calculated for the two types of plants. The correlation was very low ($r^2 = 0.08$) for plants without ants. For plants with ants, it was high ($r^2 = 0.88$) and showed a high cost of defence in young plants, which are those most vulnerable to herbivores.

Fig. 5.44. The nutrient cost relationship for defence and leaf area, in an interaction between plant, herbivore, and protective ants
(Heil et al., 1997).

rich in secondary metabolites; it therefore probably has a digestive metabolism that allows it to neutralize these compounds.

5.3.5. *Plant-animal interactions and technology*

The active ingredients of plants that have an effect against animals can be used to control predators or parasites on animals of economic value. The acarid *Varroa jacobsonii Oudermans* is particularly formidable in beehives of *Apis mellifera* L. Natural acaricides have been extensively researched and tested (Calderone and Spivak, 1995). A combination mainly made of thymol and a small amount of camphor, menthol, and eucalyptus oil, on a base material, was tested in the infested hives and resulted in a mortality of 96.7% of *Varroa*. Another essential oil extracted from a tropical plant was tested as a tick and weevil repellent (Ndungu et al., 1995). The tick (*Rhipicephalus appendiculatus Neumann*) is a vector of *Theileria parva*, which causes heavy damage in cattle of East Africa. The repellent power of the extract is equivalent to that of commercial chemical products.

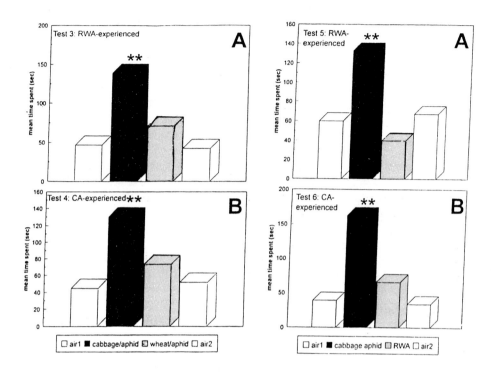

Wasps stayed longer in enclosures that attracted them. The time comparison test was that of Kruskal-Wallis,** indicating a significant difference. Air 1 and 2 = control cages with humidified air. Tests 3 and 5, wasps that had previously oviposited on wheat aphid (RWA). Tests 4 and 6, wasps that had previously oviposited on cabbage aphid. Test 3, black bars, enclosures with aphid-infested cabbage leaves. Test 4, grey bars, enclosures with aphid-infested wheat leaves. Infestations correspond to densities effectively observed in the field. Tests 5 and 6, as in the earlier tests, but aphids were removed 45 min. before wasps entered the cages. Even if there was a prior experience of oviposition on wheat aphid, wasps that could choose specifically preferred cabbage aphid, or even the odour of aphids and cabbage.

Fig. 5.45. Olfactory response of parasitoid wasp to aphid on cabbage
(Reed et al., 1995).

5.4.   Complex Interactions at the Ecosystem Level and Incidences of Evolution

5.4.1. Interactions and localization of a species

A type of interaction may lead a species to colonize a particular biotope. For example, positive plant-plant interaction (Hacker and Bertness, 1995) allows *Iva frutescens* to colonize a low salty marsh. Because of the presence of *Juncus gerardii*, the *Iva* establishes itself more easily. A plant-insect interaction may constrain a plant to live in a particular type of place. For example, *Cardamine cordifolia* is found more frequently in the shade than in the sun (Louda and Rodman, 1996). If the shade is removed above the cardamines that have been growing in the shade, they become vulnerable to herbivory and grow more slowly than plants growing continuously in the sun. However, there is no differentiation of a shade-loving ecotype that resists herbivores. If the cardamines in the sunlight are subsequently protected with insecticides, they do not grow any better than in the shade. There is no longer differentiation (independently of biotic pressures) of an ecotype that is more adapted to the shade or the sun. In the field, the plants in the sun are observed to carry many more insects than those in the shade. Curiously, cardamines in the sun have less glucosinolate, which is explained perhaps by general conditions that are less stressful than in the shade. Ultimately, it is the low herbivore pressure that explains why cardamines "take refuge" in shaded areas.

5.4.2. Seasonal effects
Seasonal effects are observed in simple plant-animal relationships. They are also found at the general level of plant-animal relationships in ecosystems that have been relatively undisturbed, for example in a small island in New Zealand (Castro and Robertson, 1997). In this situation, several plants with nectar-producing flowers are sufficient for insects as well as birds. The birds are active especially during the fairly cold seasons and on days in which pollination activity of insects is reduced. The activity of the birds compensates the reduced activity of the insects.

5.4.3. Predation and plant succession
Kollmann (1997) studied the progression of seeds on a meadow abandoned for 20 to 30 years and in the process of being overrun with ligneous species, in Central Europe. The type of production and seed dissemination was closely linked to the successional stage (Fig. 5.46). Seed dispersal and predation of grains by rodents and frequency of seeds of shrubs with fleshy fruits among the overall seed reservoir were all higher in mature shrub grasslands. The plantlets were more dense, however, at an intermediate stage in the succession. During that stage, seed fall was perhaps lower but compensated by lesser predation and lower seedling mortality due to better illumination than when the shrub cover became denser (Table 5.32). Overall, the intermediate stage represents a "window of regeneration" of shrubs with fleshy fruit, such as *Crataegus monogyna*, *Ribes rubrum*, and *Sambucus nigra*.

5.4.4. Complex interactions and evolution
There are classic examples of coevolution in plant-herbivore or host-parasite systems. More complex systems are now being better understood.

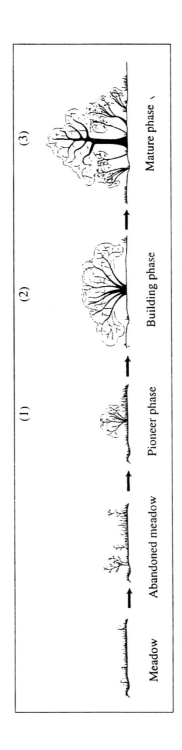

(1)   (2)   (3)

Meadow   Abandoned meadow   Pioneer phase   Building phase   Mature phase

In the pioneer phase (from 5th to 15th year), ligneous species begin to eliminate herbs. In the building phase, an intermediate shrubby stage (from 15th to 50th year), shrubs with berries regenerate the best. In the mature phase (after 50 years), there are trees.

Fig. 5.46. Phases of plant succession in Central Europe from a meadow to a forest (*Kollmann, 1997*).

Table 5.32. Seed fall and germination in an ecosystem (Kollmann, 1997)

Year	Type of habitat	N	Seedlings per quadrat		% of plants surviving	
			Mean ± SD	Median	Mean ± SD	Median
1992	Mature phase	12	8 ± 2	5 a	20 ± 6	19
	Intermediate phase	21	22 ± 4	11 b	43 ± 7	42
	Pioneer phase	3	3 ± 1	1 c	26 ± 13	33
1993	Mature phase	9	5 ± 1	3 a	7 ± 6	0 a
	Intermediate phase	15	17 ± 4	10 b	48 ± 7	50 b
	Pioneer phase	6	4 ± 1	3 a	35 ± 8	29 b

N, number of quadrats of 1 m² used for the analysis. Only quadrats with more than 5 seedlings were used. Seedlings of species of shrubs with berries were counted. The Kruskal-Wallis test revealed a significant difference between the habitats. A multiple comparison of pairs (Dunn test) was also done. Where they are different, the letters alongside indicate the significant differences ($P < 0.05$).

We have already analysed an interaction between plant, herbivore, and ant (Heil et al., 1997). In a very similar system in Southeast Asia, Federle et al. (1997) showed that trees associated with ants had deposits of lipid crystals on their stems that made them slippery, not for the symbiotic ants, but for other carnivorous ants. There had been a plant-ant coevolution that led to a highly specific association. The cuticle also protected the plant from herbivorous insects (Eigenbrode et al., 1995)

In systems involving plants, herbivorous insects, and parasitoid wasps, we also see the intervention of evolution. For a long time, it was thought that there was a specificity between chrysomelid species (herbivorous Coleoptera) and plant species. In North America, two species of chrysomelids feed on Asteraceae and Ambrosiaceae: *Ophraella notulata* feeds on *Iva frutescens*, which often forms monospecific settlements in a salty marsh, and *O. slobodniki* feeds on *Ambrosia artemisifolia*, a plant that lives in an ecosystem with numerous plant species and also carries many species of insects (Keese, 1997). The predators were more abundant on plants with *O. slobodniki*, and it was also observed that predation of eggs was greater on those plants than on plants with *O. notulata*. Emergence from eggs was not different between the two species of chrysomelids. On the other hand, the mortality of eggs by desiccation was higher in *O. notulata*, as was the parasitism of larvae, which ultimately contributed to maintain the equilibrium of the two species to some extent. Behind this appears an evolutionary problem. In the beginning, chrysomelids probably escaped from the *Ambrosia* ecosystem to another environment, that of *Iva*, which is much less rich in species, and thus also in species of predators (on eggs or on larvae). *Ophraella notulata* was trapped again by evolution because the parasitoid wasps very quickly adapted to a new environment (saline) and a new species (*Iva*). Another example involving the evolutionary rate of parasitoids can be analysed in the studies of Brown et al. (1995).

Two species of *Solidago* found in North America were attacked by a herbivorous Diptera that forms galls (recall in passing that *Solidago virga-aurea* is a classic example of ecotype). *Solidago altissima* was attacked in nearly all places. The other species, *S. gigantea*, was attacked in only one region. The Diptera were themselves attacked by a parasitoid wasp. *Solidago gigantea* is probably a new species that was followed by the Diptera, which had managed to escape on this new species. Heterozygosity was much greater in *S. altissima*, which testifies to its relative antiquity. It was observed that parasitoid wasps seldom attacked the galls of *S. gigantea*. In parallel, the new species of *Solidago* defended itself better (by chemical means) against the Diptera, which led to greater direct mortality of the Diptera. According to experimental laboratory studies, the wasp prefers to find the older plant; it thus evolved less quickly than the Diptera and the plant itself. From this we deduce that Diptera efficiently escape the parasitoid wasps. However, the Diptera itself did not effectively "follow" the plant evolution, since it attacked *S. gigantea* in only one region. We must note that other actors modulated the overall situation. Green woodpeckers attacked the galls, and there was also a preferential attack on the galls of *S. gigantea* by carnivorous Coleoptera.

5.5. Cost of Defence for the Plant

According to Coley et al. (1985), a plant can choose between two extreme positions. It may have a high growth rate (conditions of the biotope permitting) without investing in means of defence, and thus be subjected to attack from herbivores. Or it may have a low growth rate, making large investments in means of defence, and thus suffer fewer attacks from herbivores (Table 5.5). The second situation occurs in various biotopes and also in dioecious plants with female plants that need to defend their seeds.

The investment in defence is distributed between "immobile" defences such as the formation of spines and "mobile" defences, which consist essentially of secondary metabolites. The system of Coley et al. (1985) can be modelled and tested (Basey and Jenkins, 1993). For example, *Populus tremuloides*, a tree that reproduces very early (in 5 years) but lives long (200 years), is attacked especially by the beaver and produces several secondary metabolites, such as salicortine and tremulacin, which act against insects. Salicin especially is simultaneously antifungal and repulsive to beavers.

Vrieling (1991) attempted a slightly different approach. According to him, it was possible to estimate the cost of anti-herbivore defence according to the ratio between the concentration of secondary metabolites in the plant plus the mechanical defences and the reproductive effort or even vegetative growth (situations that can be observed, for example, with a biennial plant). A negative correlation was observed between growth and the concentration of secondary metabolites (essentially pyrrolidinic alkaloids) (Vrieling, 1991). Still, there was no clear relationship between concentration of secondary metabolites and mechanical resistance of leaves of *Senecio jacobaea* (which represented an investment in maintenance tissues for the vegetative apparatus) or different characters of the reproductive effort.

In laboratory conditions (Vrieling and Van Wijk, 1994a), many clones of *Senecio jacobaea* with high concentration of secondary metabolites grew more slowly than other clones with low concentration. Still, the investment calculated in secondary metabolites was low (Vrieling and Van Wijk, 1994a). In fairly limiting light (38 W/m^2), it was only 0.5% of the total organic matter, and the nitrogen deficiencies (needed for the alkaloids) and phosphorus deficiencies had no clear effects on the rates of synthesis.

In natural conditions, in this case an experimental field (Vrieling and Van Wijk, 1994b) from which herbivores could be kept out, significant relations were not observed between growth and size of the flowering or secondary metabolites. However, herbivory on the part of the Coleoptera *Longitarsus jacobaea*, by far the major actor, was negatively correlated with the concentration of secondary metabolites, which clearly shows the impact of the latter.

5.6. Mechanisms of biotic interactions

5.6.1. Introduction

Plants have evolved to defend themselves against insects, other herbivores, and pathogens. The defence is based essentially on both mechanical and chemical means. The chemical compounds involved in the interaction are mainly phenols, non-proteic amino acids, alkaloids, glycosides, terpene compounds (Table 5.5), and an entire range of peptides and glycopeptides: protease inhibitors, inhibitors of α-amylase, lectins, formidable toxins such as abrin, phenoloxidases, peroxidases, and various hydrolases such as β-glucanases and chitinases.

Plants respond to aggression in many ways, activating genes and inducing the formation of products such as terpenes (Table 5.33). The processes of activation of defence initially involve a primary stimulus that could be another organism or another form of stress.

The defences induced include physical and chemical barriers, which are stimulated by unfavourable biotic factors with more or less serious consequences as well as abiotic factors (Gomes and Xavier-Filho, 1994) (Table 5.35). There are various processes of induction. Resistance may be elicited by prior inoculation of a pathogen, following which the induction factor is transported to an induction site in various parts of the plant (systemic induction) or gives rise to a localized resistance at the site of inoculation.

Table 5.33. Content of terpenes in resins of *Pinus resinosa* Ait. (red pine) following a mechanical or fungal attack (Klepzig et al., 1995)

Monoterpene	Injury		Inoculation		F	P
	Control	Mechanical	1 week	3 weeks		
α-pinene	8.0 (7.5) a	113.5 (1.8) b	306.1 (34.7) c	288.6 (47.9) c	18.80	0.0001
camphene	0.003 (0.0) a	1.1 (0.1) b	3.1 (0.3) c	2.9 (95) c	17.24	0.0001
β-pirene	0.1 (0.0) a	38.9 (7.6) b	102.4 (17) b	118.2 (24.2) b	11.40	0.0001
ε-carene	0.01 (0.0) a	3.6 (1.0) b	9.3 (1.9) b	9.0 (2.3) b	7.02	0.0003
mycrene	0.001 (0.0) a	0.9 (0.1) b	2.5 (0.3) c	2.6 (0.5) c	13.69	0.0001
limonene	0.001 (0.0) a	1.5 (0.2) b	4.0 (0.5) c	4.0 (0.6) c	16.59	0.0001
Total	8.2 (35.6) a	159.9 (24.2) b	427.7 (50.9) c	425.5 (71.0) c	18.20	0.0001

Concentrations are in mg resin/g dry matter. Plants were inoculated with *Leptographium tenebrantis*, and the effects were observed after 1 or 3 weeks. The standard deviations are in parentheses. Different letters indicate significant differences ($P < 0.05$). F and P are determined (ANOVA) for each monoterpene per treatment, with 3 degrees of freedom.

Table 5.34. Effects of disease on natural plant populations (Jarosz and Davelos, 1995)

Disease category	Effect of infection on plant adaptation (–, negative; 0, no effect; +, positive)	Degree of pressure and incidence	Spatial variability	Temporal variability
From soil	survival (–)	locally high	very high, depending on habitat and geography	periodic cycles of the disease (e.g., *Phytophtora cinnumoni*)
Withering or ulceration	survival (–) growth (–)	could be ~ 100% (*Cryptonectria parasitica* and *Ophistoma ulmi*)	low (same species)	low (same species)
Foliar lesions	survival (– to 0) reproduction (– to 0) growth (– to 0) dynamic competitiveness (– to 0)	highly variable	very high; often linked to habitat and climatic variables	very high; linked to climatic variations
Flower and fruit diseases	reproduction (– to 0)	variable, could be very high	high, but linked to variables of habitat	very high; linked to climatic variations
Disease of whole plant	survival (–to +) reproduction (– to +) growth (– to +) competitiveness (– to +) sensitivity to other pathogens (0 to +) e.g., vegetative stimulation if mainly reproduction is affected, or even herbivores may be repelled.	variable, could be very high	low or high, often linked to characteristics of population (distributed in patches) or to beneficiaries of infection (endophytes)	

The current term used in the literature is "elicitor" to designate anything that has the property of a signal, without always distinguishing what represents a primary stimulus (of stress)—heat shock, mechanical injury by an herbivore or by the wind, UV radiation, frost, attack by the hydrolases of a fungus, and so on—from what is an alarm signal of distress, formed by the tissue attacked or by the cells of the attacking organisms, which signal will cause physical and chemical reactions in the target tissues. The term "elicitor" should be reserved for this alarm signal (a secondary stimulus), which covers in fact a large variety of chemical substances. Other chemical substances form in reaction to stimulus from these elicitors, and they constitute the defence substances strictly speaking.

5.6.2. Mechanisms of plant-fungus or plant-bacterium interactions

At the outset, the elicitors, which come from the plant or the fungus, must be distinguished from defence substances strictly speaking, which can be either essential or induced.

Certain specific substances of plants are highly antifungal, for example thymol and carvacrol, and they can be used in medicine, for example, to control fungi that affect AIDS patients (Viollon and Chaumont, 1994).

Plant resistance can be classified as follows (Heath, 1996):

- Specific resistance to a parasite. Certain clones of the plant are resistant to certain clones of a parasite species, or several species.

- Particular resistance and resistance of a cultivar. Certain cultivars of a host show resistance to all the clones of a parasite species, or several species.

- Non-host resistance. This is a partly constitutional resistance (of a plant that is not here a particular host) against a large number of potentially pathogenic species.

- Specific resistance of an organ. This resistance explains why, for example, a fungus parasitic on leaves cannot attack the roots of the same host. It seems that various fungal cutinases are necessary to help a fungus penetrate the different parts of a plant.

- Resistance linked to age. Two situations have been observed: old leaves can become more resistant or they may be attacked more at the end of the season. A seasonal effect must be taken into account, with its climatic component. When resistance increases with age, it seems to be due to progressive accumulation of substances that in fact began as basic (non-specific) defence substances. In other situations, there is only a late activation of genes.

- Induced resistance. Induced resistance is, of course, particularly interesting from the physiological point of view. It has often been remarked that chemical substances (elicitors) or certain abiotic treatments applied before the infection cause a plant known to be sensitive to develop resistance to the infection. In practice, these factors make the plant react more rapidly and effectively to an infection. Genes that were "dormant" are activated and thus produce substances of resistance that could be either very basic, or highly specific.

Induced defence systems have been studied extensively. The reader may refer to a general article on induction of defence genes (Alexander et al., 1994). Pathogens have effects at several levels, observed in cultivated and natural environments (Table 5.34). These effects vary greatly as a function of the climate.

Plant reactions range from the cellular scale to that of natural populations. For example, increase in the level of oxygenated water in cells may favour the induction of systemic resistance (Fig. 5.47). Two aspects of plant reactions particularly merit attention: formation of phytoalexins and formation of necrosed reduced zones.

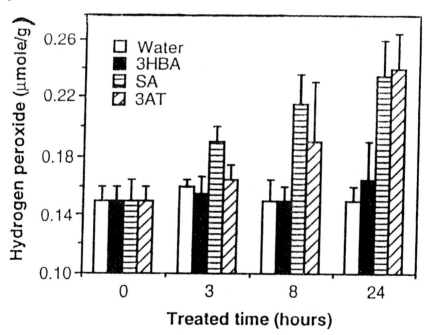

Petioles of tobacco leaves were soaked in solutions that either favoured or did not favour progressive increase in cellular oxygenated water, which plays a role in induction of systemic resistance to a large number of pathogens. 3HBA, 3-hydroxybenzoic acid (1 mM), an inactive analogue of salicylic acid. SA, salicylic acid (1 mM). 3AT, 3-amino-1,2,4-triazole, an inhibitor of catalase. The leaves remained in contact for 24 h, 14 h in light, and were then fixed according to level of H_2O_2. The standard deviations are indicated with n = 3.

Fig. 5.47. Increase in level of oxygenated water in leaf tissue and
induction of systemic resistance
(Chen et al., 1993).

Formation of phytoalexins can be induced by "abiotic elicitors" or by elicitors of biological origin. Many abiotic elicitors are known: e.g., UV at 254 nm or 345 nm, saccharose, $CuCl_2$, $CHCl_3$, salicylic acid. They do not have a general effect. For example, in sunflower, UV rays proved effective at $\lambda = 254$ nm but not at $\lambda = 345$ nm (Gutierrez et al., 1995). $CuCl_2$ is effective, but salicylic acid is not. Saccharose and salicylic acid are considered abiotic here because strictly speaking they are not elicitors in nature.

Mediation of the action of elicitors is gradually being understood. Inhibitors of the calcium/calmodulin system (Vogeli et al., 1992) can be used in tobacco. They cause the inhibition of biosynthesis of sequiterpenes, including phytoalexins of tobacco. In these conditions, the elicitor—a hydrolysate of fungal cell walls—cannot act. The calcium/calmodulin system makes the link between the elicitor and the synthesis of phytoalexins.

Defence by induction of necrosis has already been extensively studied in potato and tobacco plants. In potato, Strittmatter et al. (1995) studied interaction with *Phytophtora*. The formation of necrosed sites (comparable to those observed in hypersensitive reactions) at places attacked by *Phytophtora* caused an arrest or diminution of growth and reproduction in the fungus. Transgenic plants can be developed that express a bacterial ribonuclease, barnase, the expression of which is facilitated by ethylene. This enzyme is cytotoxic and causes necrosis. The effect of the enzyme is reduced to a limited zone by means of another transformation that allows the expression of (partial) inhibition of barnase activity. The result is a plant showing characteristic protective necrosis and/or blockage of the development of *Phytophtora*.

In tobacco, local necrosis was induced by proteic elicitins excreted by *Phytophtora* (Bonnet et al., 1996). The tobacco thus becomes resistant to this parasite. The resistance is systemic, and elicitin can be injected into the petiole of a leaf. This leaf and the others will be protected. Among these elicitins, cryptogeny shows a threshold effect at just 0.1 nmol/plant. The proteic elicitors thus studied show quite a good specificity of action in many cultivars of *Nicotiana tabacum*, but less for other species of *Nicotiana*, and no action for most other species of plants.

5.6.3. Mechanisms of plant-animal interactions

In the short term, the plant attempts to repel, dissuade, or poison a herbivore. This is a direct action. It may also attract or facilitate the intervention of a "defence auxiliary", a secondary consumer that preys on the herbivores. This is an indirect anti-herbivore action. Both types of action rely on systems of information and chemical or mechanical means.

Among the information systems, odour is generally very important in plant-animal interactions. The odour of plants plays a major role in pollination (Tollsten et al., 1994): e.g., attraction and compensation for the insects, sexual stimulation, and false copulation. Odour often acts over a great distance, and there are more or less specific plant-insect associations. Some insects are generalists.

In plant chemotaxonomy, odorous compounds are studied. It is possible to distinguish three species of angelica by PAC on 25 samples and 48 odorous compounds, although the inflorescence of Umbelliferae is easily accessible to many insects (Tollsten et al., 1994). Behind the apparently generalist interactions, there may be specific plant-insect attraction.

Odour also plays an important role in indirect defence systems. For example, the odour of cotton can attract parasitoid wasps (Rose et al., 1998). In cabbage infested with parasites, we have seen that odour attracts the wasps in experimental conditions (Reed et al., 1995).

The role of sight is also important. The significance of flower colour in pollination by many insects has been known for a long time. But sight is also involved in the browsing of plants by herbivores, and plants may be concealed by neighbouring plants (Hacker and Bertness, 1995, 1996) (Fig. 5.36).

We must also examine the possible role of mechanical information. For example, Casas et al. (1998) studied the effect of vibration of leaves and coexisting air movements on the behaviour of parasitoid wasps.

In the medium term, there is also the problem of memory, for the plant and/or the insect. A population of herbivorous insects may be observed to return preferentially to the same plant on the following year, or the contrary. Although "memory" from one year to another is sometimes observed, in other situations it seems weak or absent (Harrison, 1995).

The plant presents a wide variety of means of defence. This includes, of course, all the defence substances—toxic, repellent, dissuading—as well as anything that may attract a "defender" animal—such as nectar or even nutrient food bodies for ants (Heil et al., 1997). The plant also facilitates access to predators on the herbivores by mechanical means, as with a coating on stems and leaves that helps carnivorous insects move easily over them (Eigenbrode et al., 1995; Fig. 5.41).

Interference with the biotope has effects that cannot be easily explained. There are also effects that arise from the soils, which contain elements in levels that are useful as well as useless or even dangerous (Fig. 5.48). These effects echo all the way up the food chain (Table 5.35).

Finally, there are general mechanisms of coevolution. Plants and herbivores evolve at highly variable rates (see section 5.4.4), which can be explained at least partly by the intervention of a large number of partners in natural ecosystems.

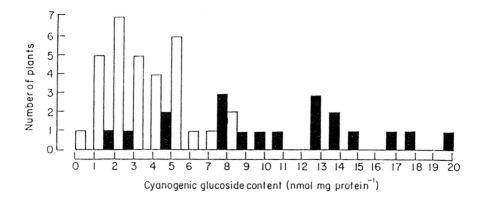

The plants were collected in England at Tooting (61 m, black bars) and at Cheviot (610 m, white bars). All the plants tested were of the Ac li genetic type, i.e., they had no linamarase, which can liberate cyanide. Only young leaves were analysed, after the plants were grown for one year all at the same place. The plants from Cheviot had low cyanogenic glucoside content, partly because many of them are heterozygous (Ac ac) for glucoside synthesis.

Fig. 5.48. Cyanogenic glucoside content of birdsfoot populations growing at two altitudes *(Hughes, 1991).*

Table 5.35. Useful elements and pollutants in a forest reserve 350 km northwest of Moscow (Markert, 1991)

	Baseline data for Central Europe	Forest Biosphere Reserve USSR
Al	400	305
Ba	5	7.8
Ca	1,000	2,710
Cd	0.2	–
Cr	3	1
Cu	12	16.6
Fe	400	165
K	7,500	15,390
Mg	500	240
Ni	15	–
Pb	10	2.7
Sr	4	4.8
Ti	8	–
Zn	40	46

The results were compared to those of a group of German forests considered by Markert (1991) to be moderately polluted. Although the values of useful elements such as Ca, Mg, and K were comparable, lead pollution (which can travel very far) was significantly lower in Russia than in Germany probably because there was less vehicular traffic.

5.6.4. Mechanisms of plant-plant interaction

In the context of plant-plant interaction, informative mechanisms are of major importance, given the immobility of plants. The synthetic review of Aphalo and Ballare (1995) particularly emphasizes the incoming signals from modifications of the environment caused by neighbouring plants. Neighbouring plants can be considered agents of stress. Another problem is acquisition and use of food resources, which depend on the use of information in the context of the **overall reaction** (adaptation, competition) of the plant.

The signals are of various kinds, including light, signals transmitted by the air or soil, and signals emitted directly or indirectly by organisms. An example is the communication for establishment of symbiosis between *Rhizobium* and a legume, or even the natural grafts that occur between the roots of individuals of the same species.

Information and action by soil
Interaction by the soil has been studied especially since the beginning of the 20th century. The works of Guyot (1991) are relevant in this context. This interaction occurs in the more general framework of physical constraints. It is a key to understanding the distribution of species in an ecosystem, as well as the distribution of a single species. The role of water is also essential in this interaction, in itself and as a vehicle of transport of allelopathic substances.

The vertical and horizontal distribution of roots is a key to competition between shrubs of the semi-desert (Briones et al., 1996). The cactus *Opuntia* is visibly affected by the presence of *Hilaria mutica* and *Larrea tridentata*, these two latter species having no effect on each other. Here, there is competition for water, but there is perhaps also an allelopathic component. The allelopathic effects could be highly specific, for example the inhibitor effect of *Artemisia annua* L., on *Amaranthus retroflexus* L., the most effective molecule being a sesquiterpene-lactone (Lydon et al., 1997).

An important problem is the **persistence** of allelopathic substances in soil, which very likely led to the first empirical observations. In this context, researchers have tested allelopathic plants that are dead but continue to be active (Bosy and Reader, 1995), for example, in the form of litter that will inhibit the germination and growth of certain adventitious plants from crops. The subject is of interest in agricultural biology.

The technical aspects of the study of these soil interactions are interesting. Observation chambers can be used to observe root growth inhibition (Mahall and Callaway, 1992) other than that resulting from simple competition for use of limited resources. For example, *Larrea tridentata* exerts a clear allelopathy on *Artemisia* (in an experiment, activated carbon was incorporated in soil to neutralize allelopathic effects). This factor may explain a fairly regular dispersal of *Larrea* in the semi-deserts of North America. In experimental interaction with *Larrea*, *Artemisia* defended itself with a better capture of water, which allowed it to maintain itself and to exploit another system of information for water level and water potential.

Several authors have used **aqueous extracts** of tissues, which are active (Chou et al., 1995), to find the active molecule in the soil where the plants producing it are already growing, for example, the p-hydroxycinnamic acid produced by *Vigna radiata*, which can affect a later culture of the same species. The effect is complex because the substances are metabolized by soil micro-organisms, the activity of which is modulated by soil pH.

In a system of hydroponic culture, Liu and Lovett (1993) observed the effects of allelopathic substances that had spread through the water of a culture and subsequently affected the growth of another hydroponic culture located at a lower level.

In studies on soil interactions, rigorous experimental protocols must be developed and results must be put through mathematical analysis. Drawing conclusions from the effects on seedlings is a complex task, given the physiological variations (e.g., dormancy, photoperiod) between species and stages of seedling growth, as well as the modes of calculation (Chiapusio et al., 1997).

Some delicate issues, such as intraspecific variation in allelopathy, can be tackled (Tarayre et al., 1995). Intraspecific variation in production can be very wide. Tarayre and colleagues experimented with great care on six chemotypes of thyme. With thyme extracts or with pure essences the variation is wide with respect to effect on *Brachypodium phenicoides*, which is a sympatric Poaceae of thyme. The effect is also always significant on seeds of a single species (thyme) but without clear distinction between the species of thyme tested. It was also observed that the inhibition of seedlings of a chemotype by the same chemotype was not different. Thus, even a chemotype relatively moderately armed with defence substances strongly inhibits the germination of its own seeds, which could have the purpose of preventing the growth of new plants too close to the mother plant, even for a chemotype that grows sparsely on the land.

Information by air and mechanical actions
Information by air involves volatile substance: e.g., ethylene (with multiple effects), isoprene, terpenes, perhaps also jasmonic acid, all substances that act as pheromones. Jasmonic acid is extensively studied. In nature it is often involved in the transmission of mechanical effects such as herbivory stress or wind stress. Jasmonic acid acts as an internal signal (equivalent to a hormone) and has allelopathic effects. It is effective when dissolved in water and thus can act through the soil. To the extent that it and its derivatives are more or less volatile, it can also act through the air.

Jasmonic acid inhibits the germination of many species of angiosperms, especially at low temperature. At moderate temperature (23°C), it acts in synergy with abscisic acid (Wilen et al., 1993), which indicates frost tolerance, among other things. Jasmonic acid also induces alterations in the supramolecular organization of thylakoids through fluorescence (Ivanov and Kicheva, 1993; Alexander et al., 1993; see Chapter 2, section 4).

For a long time it has been known that jasmonic acid induces inhibitors of proteases following mechanical stress (injury, bites by herbivores). According to Farmer et al. (1994), this induction involves an **intermediary**, an octodecanoic hydroperoxide derived from linolenic acid and leading to the formation of jasmonic acid. Jasmonic acid can be considered a distress hormone, with which ursolic acid (a triterpene) acts as an antagonist (Wasternach et al., 1994). A reduction is observed in the transcription of genes of proteins induced by jasmonic acid.

Interactions in relation with or through the effect of light
The foliage of a plant transmits and also reflects a great deal of red and far red light. Ballare et al. (1991) showed that the stem receives signals of far red light and elongates. In experiments, stems were surrounded by a copper sulphate solution allowing far red light to filter through. The plant had an escape reaction towards the top to avoid the shade created by neighbouring plants and discovered by means of far red light signals.

There is, of course, a mediation by phytochrome (see Chapter 4). Robin et al. (1994) worked on *Trifolium repens* L. and realized far red signals received by the terminal bud alone. They observed a reduction in the number of ramifications, but no particular effect on elongation. If the terminal bud is also a good receptor, it is likely that the light signals must act at several levels to be fully effective.

The plant can even **anticipate** future shade, from the time it detects a neighbouring plant that is still small (Schmitt and Wulf, 1993). Its reaction is expressed in increase in leaf size and plant size. When the population is not dense a linear correlation is observed between leaf and plant size. But when the population is dense the correlation is no longer linear and favours stem elongation, which also recalls an escape strategy.

Always in relation with the effect of light, the architecture of plants is adapted more or less to nearby plants in the context of competition for light (Tremmel and Bazzaz, 1995). Tremmel and Bazzaz studied five species of different morphology in experiments of competition in which each species in turn served as target species.

On the scale of ecosystems, of course, reactions to complex information are observed. In some situations, the complete plant succession (towards a climactic forest) is inhibited by communities of shrubs that establish themselves, as in a site near Montreal (Meilleur et al., 1994) along a high tension wire below which vegetation was cleared. Allelopathy may be suspected, as well as inhibition of germination of trees because of the shade.

Even with few species the interactions may be complex. Interaction between two *Typhas* species (Weisner, 1993) was studied on the shores of a Swedish lake over 13 years. *Typhas angustifolia* dominated, at about 1m/year, in water of 50 cm depth, but not at all near the surface. It was taller than *T. latifolia*, but the latter's wide leaves were an advantage in the very young stage. Very likely there were several effects, including light and the degree of inundation.

Conclusion

Ecophysiologists have till now concentrated on how plants can live in their sometimes difficult environments. In the footsteps of and along with ecologists and climatologists, they also study how plants influence their environments. Although researchers at first often worked on either animals or plants, or on bacteria, it increasingly seemed imperative to discover the roles and reactions of animals and plants together. In this context, it is essential to travel from the scale of the cell to that of the whole plant.

Over the last few decades, the world's population has increased tremendously. Disparity of natural resources has grown between countries, and within countries. Human activity exerts an increasingly strong pressure on the environment and thus on biodiversity. In parallel with, and to some extent linked to, the human pressure, there has been a warming of the climate in the face of which the world population must be fed now and in the long term.

Plant breeding, practised for a long time by farmers using traditional means and then by scientists, has been the focus in the effort to increase productivity. There is a great temptation to launch genetically modified (GM) plants without measuring the dangers from the perspective of dynamics of microorganism populations and the dynamics of animals and plants. The dynamics of plants (and thus of GM organisms) present a great potential for adaptation that is still not thoroughly understood in relation to the overall biodiversity.

In this context, ecophysiology has an essential role to play, at the crossroads of molecular biology, ecology, agronomy, and climatology.

Biological control may be valuable in the developed as well as developing countries to avoid the use of GM organisms. For it to be successful, plant-plant interactions and plant-animal interactions must be better understood. Ecophysiologists can use GM organisms, which are good laboratory tools, to clarify certain points of physiology, but those organisms must remain confined to the laboratory, even if only as a precaution.

In agronomy we must breed plants while respecting nature as far as possible. For example, in the present situation, as global CO_2 levels are rising, we must obtain cultivars that grow equally well with less water and fertilizer. Certain plants react to these new CO_2 conditions by synthesizing fewer proteins and requiring less nitrogen.

Always in relation with the climate, the ecophysiologist must further study the equilibrium of the plant, particularly the root physiology, in conjunction with phenomena of competition between plants and water use efficiency. The interactions of environmental factors must also be explored.

Ecophysiologists also need to probe alterations in biodiversity that cause changes in the properties of communities and ecosystems.

Studies on stress have been presented in this work, not only for the interest the subject itself holds, but also because the present climatic evolution seems to cause more frequent catastrophic events.

Finally, on the basis of diversified and numerous field experiments, holistic studies are required that are more finely adjusted to reality by means of modelling (see, for example, the work of Lambers et al., 1998).

It is to be hoped that more and more young scientists will enter the field of ecophysiology, in which a great deal remains to be explored.

Bibliography

Albers, B., Bray, D., Lewis, J., Raff, M., Roberts, K., and Watson, J.D. (1983). *Molecular Biology of the Cell*. Garland Publishing Inc., New York, London.

Alexander, D., Lawton, K., Uknes, S., Ward, E., and Ryals, J. (1994). *Defense-Related Gene Induction in Plants*. Research Triangle Park, North Carolina.

Alexander, G. Ivanov, and Kicheva, Maia I. (1993). Chlorophyll fluorescence properties of chloroplast membranes isolated from jasmonic acid-treated barley seedling. *J. Plant Physiol.*, 141: 410-414.

Ameglio, T., Daudet, F.A., Archer, P., and Ferreira, I. (1993). *Agronomie*, 13:751-759.

Amthor, J.S. (1989). *Respiration and Crop Productivity*. Springer.

Amthor, J.S. (1991). *Plant Cell Environ*. 14:13-20.

Amthor, J.S. (1993). In Wilkinson, R.E. (ed.). *Plant Environment Interactions*. Marcel Dekker, New York.

Amthor, J.S. (1995). *Ecophysiology of Photosynthesis*. In Schulze, E.D., and Caldwell, M.M. (1995). *Ecophysiology of Photosynthesis*. Springer, chapter 3.

Amthor, J.S., Koch, G.W., and Bloom, A.J. (1992). *Plant Physiol*. 98:757-759.

Amzallag, G.N. (1994). *New Phytol*. 128:715-723.

Amzallag, G.N. (1996). *New Phytol*. 132(2):317-325.

Amzallag, G.N., and Lerner, H.R. (1995). Physiological adaptation of plants to environmental stresses. In Pessarakli, M. (ed.), *Handbook of Plant and Crop Physiology*. Marcel Dekker, New York, pp. 557-576.

Anaya, A.L., Sabourin, D.J., Hernandez-Bautista, B.E., and Mendez, I. (1995). *J. Chem. Ecol.*, 21(8):1085-1102.

Andriolo, J.L., Le Bot, J., Gary, C., Sappe, G., Orlando, P., Brunel, B., and Sarrouy, C. (1996). *J. Plant Nutr.*, 19:1441-1462.

Angelopoulos, K., Dichio, B., and Xiloyarmis, C. (1996). *J. Exp. Botany*, 47(301):1093-1100.

Angelov, M.N., Jindong, S., Byrd, G.T., Brown, R.H., and Black, C.C. (1993). *Photosynth. Res.*, 38:61-72.

Aphalo, P.J., and Ballare, C.L. (1995). On the importance of information-acquiring systems in plant-plant interaction. *Funct. Ecol.*, 9:5-14.

Arndt, U., Nobel, W., and Schweiber, B. (1987). *Bioindikatoren Moglichkeiken, Grenzen und Neue Erkenntnisse*. Ulmer, Stuttgart. b, pp. 314-326.

Bagnall, D.J. (1992). *Aust. J. Plant Physiol.*, 19:401-409.

Ballare, C.L., Scopel, A.L., and Sanchez, R.A. (1991). On the opportunity cost of the photosynthate invested in stem elongation reactions mediated by phytochrome. *Oecologia*, 86:561-567.

Bannister, P. (1976). *Introduction to Physiological Plant Ecology*. Blackwell, Oxford.

Bannister, P., and Smith, P.J.M. (1983). *Flore*, 173:399-414.

Basey, J.M., and Jenkins, H. (1993). Production of chemical defenses in relation to plant growth rate. *Oikos*, 68:323-328.

Bauer, H. (1972). *Photosynthetica*, pp. 424-434.

Bauer, H., Nagele, M., Comploj, M., Galler, V., Mair, M., and Unterpertinger, E. (1994). *Physiologia plantarum*, 91:403-412.

Bauer, W. (1995). In Schulze, E.D., and Caldwell, M.M. (1995). *Ecophysiology of Photosynthesis*. Springer, chapter 13.

Baumeister, W., and Ernst, W. (1978). *Mineral Stoffe und Pflanzen Wachstum* 3, Aufl. Fischer, Stuttgart.

Beck, E., Schulze, E.D., Senser, M., and Scheibe, R. (1984). *Planta*, 162:276-282.

Beck, E., Senser, M., Scheibe, R., Steiger, H.M., and Pongratz, P. (1982). *Plant Cell Environ.*, 5:215-222.

Begg, J.E., and Torsell, B.W.R. (1974). *R. Soc. NZ Bull.*, 12:277-283.

Bennet, A.F. (1987). New directions in ecological physiology. In Feder, M.E., Bennet, A.F., Burgreen, W.W., and Huey, R.B. (eds.). Cambridge University Press.

Bennet, M.H., Gallagher, M.D.S., Bestwick, C.S., Rossitier, J.T., and Mansfield, J.W. (1994). The phytoalexin response of lettuce to challenge by *Botrytis cinerea*, *Bremia lactucae* and *Pseudomonas syringae* pv. *phaseolicola*. *Physiol. Mol. Plant. Pathol.*, 44:321-333.

Berger, A. (1994). *Le Climat de la Terre*. De Boeck University, Brussels.

Berger, W. (1931). *Beih. Bot. Central*, 48:364-390.

Bernard, N. (1955). *Yersin, Pionnier, Savant, Explorateur*. La Colombe, Paris.

Berry, J.A., and Raison, J.K. (1981). Response of macrophytes to temperature. In Lange, O.L., Nobel., P.S., Osmond, C.B., and Ziegler, H. (eds.). *Encyclopedia of Plant Physiology*, vol. 12A. Springer, Berlin, Heidelberg, New York, pp. 277-338.

Beyschlag, W., Lange, O.L., and Tendunen, J.D. (1990). *Flora*, 184:271-289.

Bierzychudek, P. (1982). *New Phytol.*, 90:757-776.

Bilger, W., and Bjorkman, O. (1991). *Planta*, 184:226-234.

Billings, W.D., Godfrey, P.J., Chabot, B.F., and Bourque, D.P. (1971). *Arct. Alp. Res.*, 3:277-290.

Birot, P. (1964). *Les Formations Vegetales du Globe*. Societe d'edition d'enseignement superieur.

Bjorkman, O. (1968). *Physiol. Plant*, 21:84-99.

Bjorkman, O., and Demming-Adams, B. (1995). Ecophysiology of photosynthesis. In Schulze, E.D., and Caldwell, M.M. (1995). *Ecophysiology of Photosynthesis*. Springer, chapter 2.

Bjorkman, O., Gauhl, E., Hiesey, W.M., Nicholson, F., and Nobs, M.A. (1969). *Carnegie Trust Year Book*, 69:477-479.

Bjorkman, O., Pearcy, R.W., Harrison, A.T., and Mooney, H.A. (1972). *Science*, 175:786-789.

Bjorkman, O., and Schafer, C. (1989). *Philos. Trans. R. Soc. London*, B 323:309-311.

Blanc, P. (1992). *Rev. Ecol. "Terre Vie"*, 47:3-49.

Block, W. (1991). *Funct. Ecol.*, 5:284-290.

Block, W., and Vannier, G. (1994). *Acta Oecologica*, 15(1):5-12.

Bloemen, M.L., Markert, B., and Lieth, H. (1995). *Sci. Total Environ.*, 166:137-148.

Blumenthal, M.J., Aston, S.C., and Pearson, C.J. (1996). *Aust. J. Agric. Res.* 47:1119-1130.

Bobrouskaya, N.I. (1985). *Feddes Report*, 96:425-432.

Bodner, M., and Beck, E. (1987). *Oecologia*, 72:366-371.

Bonnet, P., Bourdon, E., Ponchet, M., Blein, J.P., and Ricci, P. (1996). Acquired resistance triggered by elicitins in tobacco and other plants. *Eur. J. Plant Pathol.* 102:181-192.

Bosy, J.L., and Reader, R.J. (1995). Mechanisms underlying the suppression of forb seedling emergence by grass (*Poa pratensis*) litter. *Funct. Ecol.*, 9:635-639.

Boyer, J.S., and Bowen, B.L. (1970). *Plant Physiol.* 46:233-235.

Brenckmann, F. (1997). *Grains de Vie*. Arthaud.

Briones, D., Montana, C., and Ezcurra, E. (1996). *J. Veg. Sci.*, 7(3):453-460.

Brock, T.D. (1978). *Thermophilic Microorganisms and Life at High Temperatures*. Springer, Berlin, Heidelberg, New York.

Broekaert, W.F., Franky, R., Terras, G., Cammue, B.P.A., and Osborn, W. (1995). Plant defensins: Novel antimicrobial peptides as components of the host defense system 1. *Plant Physiol.*, 108:1353-1358.

Brooks, A., and Farquhar, G.D. (1985). *Planta*, 165:397-406.

Brown, J.M., Warren, G., Abrahamson, R., Packer, A., and Way, P.A. (1995). The role of natural enemy escape in a gallmaker host-plant shift. *Oecologia*, 104:52-60.

Brown, S.L., Chaney, R.L., Angle, J.S., and Baker, A.J.M. (1995). *Soil Sci. Soc. Amer. J.*, 59:125-133.

Brune, A., Urbach, W., and Dietz, K.J. (1995). *New Phytol.*, 129(3):403-409.

Brunig, E.F. (1987). *Ambio*, 16:67-72.

Byrd, G.T., Sage, R.F., and Brown, R.H. (1992). *Plant Physiol.*, 100:191-198.

Calderone, N.W., and Spivak, M. (1995). Plant extracts for control of the parasitic mite *Varroa jacobsoni* (Acari: Varroidae) in colonies of the western honey bee (Hymenoptera: Apidae). *J. Econ. Entomol.*, 88(5):1211-1215.

Caldwell, M.M. Meister, H.P., Tenhunen, J.D., and Lange, O.L. (1986). Trees 1. In Schulze, E.D., and Caldwell, M.M. (1995). *Ecophysiology of Photosynthesis*. Springer, chapter 20, pp. 25-41.

Caldwell, M.M., White, R.S., Moore, R.T., and Camp, L.B. (1977). *Oecologia*, 29:275-300.

Calow, P. (1987). *Funct. Ecol.*, 1:57-61.

Cape and Vogt (1991). *Can. J. For. Res.*, 21:1423-1429.

Carter, D.L. (1982). Salinity and plant productivity. In Rechcigl, M. (ed.). *CRC Handbook of Agricultural Productivity*, vol. 1, *Plant Productivity*. CRC Press, Boca Raton, pp. 117-133.

Casas, J., Bacher, S., Tautz, J., Meyhofer, R., and Pierre, D. (1998). *Biol. Control.*, 11(2):147-153.

Caspar, T., Huber, S.C., and Sommerville, C. (1986). *Plant Physiol.*, 76:1-7.

Castro, I., and Robertson, A.W. (1997). *NZ J. Ecol.*, 21(2):169-179.

Ceulemans, R., and Saugier, B. (1991). Photosynthesis. In Raghavendra, A.S. (ed.). *Physiology of Trees*. Wiley, London, pp. 21-50.

Chartier, P.H., and Bethenod, O. (1977). La productivite primaire a l'echelle de la feuille. In Moyse, A. *Les Processus de la Produciton Vegetale Primaire*. Gauthier-villars, Paris, pp. 77-112.

Chen, Z., Silva, H., and Klessig, D.F. (1993). *Science*, 262:1883-1886.

Chiapusio, G., Sanchez, A.M., Reigosa, M.J., Gonzalez, L., and Pelissier, F. (1997). *J. Chem. Ecol.*, 23(11):2445-2453.

Chiche, J., Bekhtiar, R., and Leclerc, J.P. (1998). *J. de Botanique*, 7:53-66.

Chou, C.H., Waller, G.R., Cheng, C.S., Yang, C.F., and Kim, O. (1995). *Bot. Bull. Acad. Sin.*, 36:9-18.

Claes, B., Dekeyser, R., Villaroel, R., Van Den Buleke, M., Van Montagu, M., and Caplan, A. (1990). *Plant Cell*, 2:19-27.

Clarke, A. (1991). *Amer. Zool.*, 31:81-92.

Coley, P.D., Bryant, J.P., and Chapin, F.S. III Science, 230: 895-899.

Darral, N.M., and Jager, H.J. (1984). Biochemical diagnostic tests for the effect of air pollution on plants. In Koziol, M.J., and Whatley, F.R. (eds.), *Gaseous Air Pollutants and Plant Metabolism*. Butterworths, London, pp. 333-349.

Dassler, H.G. (1991). *Ursacher Wirkunger Gegenma ßnahmen*, 4, Aufl. Fischer, Jena.

De Filippis, L.F., Hampp, R., and Ziegler, H. (1981). *Z. Pflanzenphysiol.*, 103:1-7.

De Lima, M., and Copeland, L. (1994). *Aust. J. Plant Physiol.*, 21:85-94.

Demming-Adams, B., and Adams, W.W., III (1996). *Planta*, 198:460-470.

Deruelle, S. (1979). Doc. Thes. in Natural Sciences, University of Paris.

Dixon, R.K., Brown, S., Houghton, R.A., Solomon, A.M., Trexler, M.C., and Wisniewski, J. (1994). *Science*, 263:185-190.

Doley, D. (1981). Tropical and subtropical forests and woodlands. In Kozlowski, T.T. (ed.). *Water Deficits and Plant Growth*, vol. VI. *Woody Plant Communities*. Academic Press, New York, pp. 109-233.

Dolman, A.J., Gash, J.H.C., Roberts, J., and Shutleworth, W.J. (1991). In Schulze, E.D., and Caldwell, M.M. (eds.) (1995). *Ecophysiology of Photosynthesis.* Springer.

Douce, R., and Neuburger, M. (1989). *Annu. Rev. Plant Physiol. Plant Mol. Biol.,* 40:371-414.

Downs, R.J., Nadelhoffer, K.J., Melillo, J.M., and Aber, J.D. (1993). *Trees,* 7:233-236.

Dreux, P. (1986). *Precis d'Ecologie,* 3d ed. PUF, Paris, 282 pp.

Dufrene, E., and Saugier, B. (1993). *Funct. Ecol.,* 7:97-104.

Durand, P., Neal, C., Lelong, F., and Didon-Lescot, J.F. (1992). *Sci. Total Environ.,* 119:191-209.

Duvigneaud, P. (1967). *Ecosystemes et Biosphere.* Min. Educ. Nat. Cult., Brussels.

Duvigneaud, P., and Denaeyer-De Smet, S. (1973). *Deco Plant,* 8: 219-246.

Dyer, L.A., and Bowers, M.P. (1996). *J. Chem. Ecol.,* 22(8):1527-1539.

Ehleringer, J.A., and Bjorkman, O. (1978). *Oecologia,* 36:151-162.

Eigenbrode, S.D., Moodies, S., and Castagnola, T. (1995). *Entomol. Exp. Appl.,* 77:335-342.

Eigenbrode, S.D., and Pillai, S.K. (1998). *J. Chem. Ecol.,* 24(10):1611-1627.

Ellenberg, H. (1986). *Vegetation Mitteleuropas mit den Alpen,* 4. Aufl. Ulmer, Stuttgart.

Epron, D. (1997). *J. Exp. Bot.,* 48(315):1835-1841.

Ernst, W.H.O. (1976). Physiological and biochemical aspects of metal tolerance. In Mansfield, I.A. (ed.). *Effects of Air Pollutants on Plants.* Cambridge University Press, Cambridge, pp. 115-133.

Ernst, W.H.O. (1990). Mine vegetation in Europe. In Shaw, A.J. (ed.). *Heavy Metal Tolerance in Plants: Evolutionary Aspects.* CRC Press, Boca Raton, pp. 21-37.

Ernst, W.H.O., and Joose Van Damme, E.N.G. (1983). *Umveltbelastung durch Mineralstoffe-biologische Effekke.* Fischer, Stuttgart.

Estienne, P., and Godard, J.A. (1979). *Climatologie.* Armand Colin.

Farmer, E., Calderadi, D., Pearce, G., Walker-Simmons, M.K., and Ryan, C.A. (1994). Diethyldithiocarbamic acid inhibits the octadecanoid signaling pathway for the wound induction of proteinase inhibitors in tomato leaves. *Plant Physiol.,* 106:337-342.

Farrant, J.M. (2000). *Plant Ecol.,* 151:29-39.

Farrar, J.F., and Williams, J.H.H. (1991). In Emrd, M.J. (ed.). *Compartmentation of Plant Metabolism in Non-photosynthetic Tissues.* Cambridge Univ. Press.

Favarger, C.L., and Robert, P.A. (1962). *Flore et Vegetation des Alpes 1, Etage Alpin,* Delaichairs et Niestle, Neufchatel.

Federle, W., Maschwitz, W., Fiala, B., Riederer, M., and Holldobler, B. (1997). *Oecologia,* 112(2):217-224.

Feder, M.E. (1987). In Feder, M.E., Bennet, A.F., Burgreen, W.W., and Huey, R.B. (eds.). *New Directions in Ecological Physiology.* Cambridge Univ. Press.

Ferder, M.E., and Black, B.A. (1991). *Funct. Ecol.,* 5:136-144.

Fichtner, J., Koch, G.W., and Mooney, H.A. (1995). In Schulze, E.D., and Caldwell, M.M. (1995). *Ecophysiology of Photosynthesis.* Springer, chapter 7.

Fitzpatrick, J., and Kirkman, H. (1995). *Mar. Ecol. Prog. Ser.,* 127(1-3):279-289.

Flowers, I.J. (1985). *Plant Soil,* 89:41-56.

Frederick, S.E., and Newcomb, E. (1969). *J. Cell. Biol.,* 43:343-353.

Frederick, S.E., and Newcomb, E. (1971). *Planta,* 96:152-174.

Frohne, D., and Jensen, U. (1992). *Systematik des Pflanzenreichs.* Fischer Verlag, Stuttgart.

Fujikawa, S., and Miura, K. (1986). *Jpn. J. Freeze Dry,* 32:14-17.

Gaff, D.F. (1980). Photoplasmic tolerance of extreme water stress. In Turner, N.C., and Kramer, P.J. (eds.). *Adaptation of Plants to Water and High Temperature Stress.* Wiley, New York, pp. 207-230.

Galsomies, L., Robert, M., Gelie, B., and Jaunet, A.M. (1992). *Bull. Soc. Bot.,* 139(2):25-32.

Gamon, J.A., Penvelas, J., and Field, C.B. (1992). *Remote Sens. Environ.,* 41:35-44.

Gebauer, G., Rehder, H.J., and Wollenweber, B. (1988). *Oecologia,* 75:371-385.

Gein, A.M. (1972). *Umschau,* 72:551-554.

Genty, B., Briantais, J.M., and Baker, N. (1989). *Biochim. Biophys. Acta,* 990:87-92.

Getz, H.P. (1991). *Planta,* 185:261-268.

Giacolou, J., and Echaubard, M. (1987). *Hydrobiologia,* 148:269-280.

Gimmler, W., Weis, U., and Weiss, C. (1989). pH-regulation and membrane potential of extremely acid-resistant algae. In Dainty, J., De Michaelis, M., Marré, E., and Rasi-Caldogno, F. (eds.), *Plant Membrane Transport.* Elsevier, Amsterdam, pp. 389-390.

Gojon, A., Bussi, C., Grignon, C., and Salsac, L. (1991). *Physiol. Plant,* 82:505-512.

Goldstein, G., Rada, F., and Azocar, A. (1985). *Oecologia,* 68:147-152.

Golley, F.B., McGinnis, J.T., Clements, R.G., Child, G.I., and Duever, J.M. (1975). *Mineral Cycling in a Tropical Moist Forest Ecosystem.* University of Georgia Press, Athens.

Gomes, V.M., and Xavier-Filho, J.X. (1994). Biochemical defenses of plants. *Arq. Biol. Technol.,* 37(2):371-383.

Grace, J. (1977). *Plant Response to Wind.* Academic Press, London.

Grill, D., Guttenberger, H., Zellning, G., and Bermadinger, E. (1989). *Phyton,* 29:277-290.

Grime, J.P., and Hodgson, J.G. (1969). *Ecological Aspects of the Mineral Nutrition of Plants*. Blackwell, Oxford, pp. 67-99.

Gruszecki, W.I. (1995). *Acta Physiol. Plantarum*, 17(2):145-152.

Gutierrez, M.C., Parry, A., Manuel, T., Jorrin, J., and Edwards, R. (1995). Abiotic elicitation of coumarin phytoalexins in sunflowers. *Phytochemistry*, 3815:1185-1191.

Guyot, L. (1991). *C.R. de l'Academie d'Agriculture de France*, 373:1-15.

Guyot, L., Guillemet, J., and Montegut, J. (1995). *Ann. des Epiphytes*, 2:119-163.

Hacker, S., and Bertness, L. (1995). *Ecology*, 7617:2165-2175.

Hacker, S., and Bertness, L. (1996). *Amer. Natural.* 148(3):559-575.

Halbwachs, G., and Wimmer, R. (1987). Halzanatomische Aspekte beider Einwirkung von Immissionen auf Bäume. In Rossmanith, H.P. (ed.)., Waldschäden-Holzwirtschaft. Österr Agrar-Verlag, WIEN, pp. 133-147.

Hall, S.C., Gang, D.R., and Weber, D.J. (1994). Seasonal variation in volatile secondary compounds of *Chrysothamnus nauseosus* (PALLAS) Britt., Asteraceae ssp. hololeucus (Gray) Hall. & Clem. influences herbivory. *J. Chem. Ecol.*, 20(8):2055-2063.

Hansson, K.R. (1992). *Plant Physiol.*, pp. 276-283.

Hansson, L. (1992). *T.R.E.E.*, 7(9):299-302.

Harborne, J.B. (1988). *Introduction to Ecological Biochemistry*, 3d ed. Academic Press, London.

Harrison, S. (1995). *Oecologia*, 103:343-348.

Hartel, O. (1976). *Umschau*, 76:347-350.

Hattersley, P.W. (1983). *Oecologia* 57:113-128.

Havaux, M. (1993). *Plant Cell Environ.*, 16:461-467.

Havaux, M. (1996). *Photosynth. Res.*, 47:85-97.

Heath, M.C. (1996). *Can. J. Plant Pathol.*, 18(4):469-475.

Heil, M., Fiala, B., Eduard Linsenmair, K., Zotz, G., Menke, P., Maschwitz, U. (1997). Food body production in *Macaranga tribola* (Euphorbiacea): A plant investment in anti-herbivore defence via symbiotic and partners. *J. Ecol.*, 85:847-861.

Heimer, Y.M. (1973). *Planta*, 113:279-281.

Henson, I.E., and Turner, N.C. (1991). *New Phytol.*, 117:529-535.

Hesketh, J., and Baker, D. (1967). *Crop Sci.*, 7:285-293.

Hoflacher, H., and Bauer, H. (1982). *Physiol. Plant*, 56:177-182.

Hofler, K., Migsch, H., and Rottenburg, W. (1941). *Forschungdienst*, 12:50-61.

Horsman, D.C., and Wellburn, A.R. (1976). Guide to the metabolic and biochemical effects of air pollutants on higher plants. In Mansfield, T.A. (ed.), *Effect of Air Pollutants on Plants*. Cambridge Univ. Press, Cambridge, pp. 185-199.

Hsiao, T.H.C., O'toole, J.C., Yambao, E.B., and Turner, N.C. (1984). *Plant Physiol.*, 75:338-341.

Huber, B. (1956). In Ruhland, W. (ed.). *Handbuch der Pflanzenphysiologie*, vol. 3. Springer, Berlin, Heidelberg, New York, pp. 509-513.

Hughes, M.A. (1991). *Heredity*, 66:105-115.

Huner, P.A., Oquist, G., Hurry, V.M., Krol, M., Falk, S., and Griffith, M. (1993). *Photosynth. Res.*, 37:19-39.

Indergit, T., and Mallik, A.U. (1997). *Plant Ecol.*, 133(1): 9-36.

Irigoyen, J.J., De Juan, J.P., and Sanchez-Diaz, M. (1996). *New Phytol.*, 134(1):53-59.

Ivanov, A.G., and Kicheva, M. (1993). *J. Plant Physiol.*, 141(4):410-414.

Jager, H.J. (1982). Biochemical indication of an effect of air pollution on plants. In Steubing, L., and Jager, H.J. (eds). *Monitoring of Air Pollutants by Plants*. Jung, The Hague, pp. 99-107.

Jarosz, A.M., and Davelos, A.L. (1995). Tansley Review no. 8. *New Phytol.*, 129:371-387.

Jeffrey, D.W. (1987). *Soil-Plant Relationship. An Ecological Approach*. Croom Helm, London.

Jeschke, W.D., Atkins, C.A., and Pate, J.S. (1985). *J. Plant Physiol.*, 117:319-330.

Johnson, M.K., Johnson, E.J., MacElroy, R.D., Speer, H.L., and Bruff B.S. (1968). *J. Bacteriol.*, 95:1461-1468.

Joyeux, M., Lobstein, A., Anton, R., and Mortier, F. (1995). *Planta Medica*, 61:126-129.

Kaiser, W.M., Kaiser, G., Martinoia, E., and Heber, U. (1988). In Kleinkauf, H., Dohren, R.V., and Jaenicke, L. (eds.). *The Roots of Modern Biochemistry*. De Gruyter, Berlin, pp. 722-733.

Kakegawa, K., Hattori, E., Koike, K., and Tadeka, K. (1991). *Phytochem.*, 30(7):2271-2273.

Kallio, P., and Veum, A.K. (1975). analysis of precipitation at Fennoscandian tundra sites. In Wielogalski, F.E., Kallio, P., and Rosswall, T. (eds.). *Fennoscandian Tundra Ecosystems*, 1, *Plants and Microorganisms*. Springer, Berlin, Heidelberg, New York, pp. 333-338.

Kappen, L. (1981). Ecological significance of resistance to high temperature. In Lange, O.L., Nobel, P.S., Osmond, C.B., and Ziegler, H. (eds.). *Encyclopedia of Plant Physiology IZA*. Springer, Berlin, Heidelberg, New York, pp. 439-474.

Kappen, L., Schultz, G., and Vanselow, R. (1995). Direct observations of stomatal movements. In Schulze, E.D., and Caldwell, M.M. (eds.), *Ecophysiology of Photosynthesis*. Springer-Verlag, Berlin, pp. 231-246.

Katz, A. (1990). Detection of diterpenoid alkaloids in aphids feeding on *Aconitum napellus* and *Aconium paniculatum*. *J. Natur. Prod.*, 53(1):204-206.

Kaul, K., and Bedi, S. (1995). Allelopathic influence of *Tagetes* species on germination and seedling growth of radish (*Raphanus sativus*) and lettuce (*Lactuca sativa*). *J. Agric. Sci.*, 65(8):599-601.

Keck, R.W., and Boyer, J.S. (1974). *Plant Physiol.*, 53:474-479.

Keese, M.C. (1997). *Oecologia*, 112(1):81-86.

Keller, Th. (1980). *Can. J. For. Res.*, 10:1-6.

Kershaw, K.A. (1985). *Physiological Ecology of Lichens.* Cambridge Univ. Press, Cambridge.

Kirschbaum, M.U.F., and Farquhar, G.D. (1987). *Plant Physiol.*, 83:1032-1036.

Klein, J.C. (1992). Doct. thesis in natural sciences. University of Paris-South.

Klepzig, K.D., Kruger, E.L., Smalley, E.B., and Raffa, K.F. (1995). *J. Chem. Ecol.*, 21(5):601-626.

Kok, B. (1948). *Enzymologia*, 13:1-56.

Kollmann, J. (1997). Regeneration window for flesh-fruited plants during scrub development on abandoned grassland 1. *Ecoscience*, 2(3):213-222.

Korner, C.H. (1995). In Schulze, E.D., and Caldwell, M.M. (1995). *Ecophysiology of Photosynthesis.* Springer.

Korner, C., and Caldwell, M.M. (1995). In Schulze, E.D., and Caldwell, M.M. (1995). *Ecophysiology of Photosynthesis.* Springer, chapter 22.

Kowallik, W. (1982). *Annu. Rev. Plant Physiol.*, 95:1270-1276.

Kowalski, S. (1987). Mycotrophy of trees in converted stands remaining under strong pressure of industrial pollution. *Angew Bot.*, 61:65-83.

Krause, G.H., and Weiss, E. (1991). *Annu. Rev. Plant Physiol.*, 42:313-349.

Krieger, A. (1992). Thesis, Heinrich Heine University, Dusseldorf.

Kromer, S., and Heldt, H.W. (1991). *Biochim. Biophys. Acta*, 1057:42-50.

Krunenberg, G.H.M., and Kendrick, R.E. (1986). The physiology of action. In Kendrick, R.E., Kronenberg, G.H.M. (eds.). *Photomorphogenesis in Plants.* Nijhoff, Dordrecht, pp. 99-114.

Krupa, Z., Baranowska, M., and Orzol, D. (1996). *Acta Physiol. Plantarum*, 18(2):147-151.

Krupa, Z., Siedlecza, A., Maksymiec, W., and Baszynki, T. (1993). *J. Plant Physiol.*, 142:664-668.

Krusmann, G. (1970). *Taschenbuch der Geholzverwenduny* 2, Aufl. Parey, Berlin.

Kuntze, H., Niemann, J., Roeschmann, G., and Schweerdtfeger, G. (1988). *Bodenkunde* 4, Aufl. Ulmer, Stuttgart.

Kutscher, L., and Lichtenegger, E. (1992). *Warzelatlas mitteleuropaischer Grunland planzen* vol. II. *Pteridophyta und Dicotyledonae (Magnoliopsida) Teil 1: Morphologie, Anatomie, Okologie, Verbreitung, Soziologie, Wirtschaft.* Fischer, Stuttgart.

Lachaud, S. (1989). *Trees*, 3:125-137.

Lajko, F., Kadioglu, A., Borbely, G., and Garab, G. (1997). *Photosynthetica*, 33(2):217-226.

Lambers, H., Stuart Chapin, F., and Thijs, L. Pons (1998). *Plant Physiological Ecology.* Springer Verlag, Berlin, Heidelberg.

Landolt, E. (1971). *Boissiera*, 19:129-148.

Lange, O.L. (1989). *J. Ecol.*, 76:915-937.

Lange, O.L., Beyschlag, W., and Tenhunen, J.D. (1987). Control of leaf carbon assimilation. In Schulze, E.D., and Zwolfer, H. (eds.). *Potentials and Limitations of Ecosystem Analysis*. Springer, Berlin, Heidelberg, New York, pp. 149-163.

Larcher, W. (1961). *Planta*, 56:575-606.

Larcher, W. (1963). *Planta*, 60:339-343.

Larcher, W. (1973). Limiting temperature for life functions in plants. In Precht, H., Christophersen, J., Hensel, H., and Larcher, W. (eds.). *Temperature and Life*, 2nd ed. Springer, Berlin, Heidelberg, New York.

Larcher, W. (1973). *Okologie der Pflanzen*, 1, Aufl. Ulmer, Stuttgart.

Larcher, W. (1994). Photosynthesis as a tool for indicating temperature stress events. In Schulze, E.D., and Caldwell, M.M. (eds.), *Ecophysiology of Photosynthesis*. Springer, Berlin, Heidelberg, New York, pp. 261-277.

Larcher, W. (1995). *Physiological Plant Ecology*. Springer.

Larcher, W., Holzner, M., and Pichler, J. (1989). *Flora*, 183:115-131.

Lauchli, A. (1976). Symplastic transport and ion release to the xylem. In Wardlaw, I.F., and Passioura, J.B. (eds.). *Transport and Transfer Processes in Plants*. Academic Press, New York, pp. 102-112.

Laver, M.J., Pallardy, S.T.G., Blevins, D.G., and Randall, D.D. (1989). *Plant Physiol.*, 91:848-854.

Lea, P.J., and Miflin, B.J. (1974). *Nature*, 251:614-616.

Leclerc, J.C. (1964). Contribution a l'etude de la localisation du chlore et du sodium chez les halophytes. DES, University of Caen.

Leclerc, J.C. (1976). Thesis, University of Paris-South.

Leclerc, J.C., and Abd el Rahman, N. (1988). *Cpt Rendu Acad. Sci. Paris Serie III*, 306:421-426.

Leclerc, J.C., and Blaise, S. (1991). *Vegetatio*, 92:85-93.

Leclerc, J.C., Coute, A., and Hoarau, J. (1981). *Photosynthesis*, VI, pp. 443-444 (ed. G. Akoyounoglov). Balaban, Philadelphia.

Leclerc, J.C., Briane, J.P., and Blaise, S. (1998). Behaviour under different light conditions of two *Arabidopsis* ecotypes. In Garab, G. (ed.), *Photosynthesis: Mechanisms and Effects*, Vol. VI, pp. 2601-2604.

Lenoble, M.E., Blevins, D.G., Sharp, R.E., and Cumbie, B.G. (1996). *Plant, Cell. Environ.*, 19:1132-1142.

Levitt, J. (1956). *The Hardiness in Plants*. Amer. Soc. Agron.

Levitt, J. (1980). *Responses of Plants to Environmental Stress*. Academic Press, New York.

Liang, N., Maruyama, K., and Huang, Y. (1995). *Photosynthetica*, 31(4):529-539.

Lichtenthaler, H.K., Buschmann, C., Doll, M., Fietz, H.J., Bach, T., Kozel, U., Meier, D., and Rahmsdorf, U. (1981). *Photosynth. Res.*, 2:115-141.

Liese, W., Schneider, M., and Eckstein, D. (1975). *Eur. J. For. Pathol.*, 5:152-161.

Lieth, H. (1975). Modelling of the primary production of the world. In Lieth, H., and Whittaker, R. (eds.). *Primary Production in the Biosphere.* Springer, Berlin, Heidelberg, New York, pp. 119-129.

Likens, G.E., Bormann, F.H., Pierce, R.S., Eaton, J.S., and Johnson, N.M. (1977). *Biogeochemistry of a Forested Ecosystem.* Springer, Berlin, Heidelberg, New York.

Lindberg, S.E., Lovett, G.M., Richter, D.D., and Johnson, D.W. (1986). *Science*, 231:141-145.

Liu,, D.L., and Lovett, J.V. (1993). Biologically active secondary metabolites of barley. I. Developing techniques and assessing allelopathy in barley. *J. Chem. Ecol.*, 19(10): 2217-2229

Lois, R. (1994). *Planta*, 194:498-503.

Lois, R., and Buchanan, B.B. (1994). *Planta*, 194:504-509.

Lockhart, J.A. (1965). *J. Theor. Biol.*, 8:264-275.

Losch, R. (1990). In *Analysis of Water Transportation and Cavitation of Xylem Conduits.* In. Workshop, 29-31 May 1990, Vallombrosa, Florence.

Louda, S.M., and Rodman, J.E. (1996). Insect herbivory as a major factor in the shade distribution of a native crucifer (*Cardamine cordifolia* A. Gray, bittercress). *J. Ecol.*, 84:229-237.

Luttge, U. (1973). *Stoff Transpoort der Planzen.* Springer, Berlin, Heidelberg, New York.

Luttge, U., Kluge, M., and Bauer, G. (1992). *Botanique.* Tec & Doc, Lavoisier.

Lutz, C. (1996). *J. Plant Physiol.*, 148:120-128.

Lydon, J., Jeasdale, J.R., and Chen, P.K. (1997). *Weed Sci.*, 45(6):807-811.

Macias, F.A., Simonet, A.M., and Esteban, D. (1994). *Phytochemistry*, 36(6):1369-1379.

Mahall, B.E., and Callaway, R.M. (1992). Root communication mechanisms and intracommunity distribution of two Mojave desert shrubs 1. *Ecology*, 73(6), Ecological Society of America.

Mallik, A.U. (1995). *Environ. Mgmt*, 19(5):675-684.

Mallik, A.U. (1995). *Forest Ecol. Mgmt*, 81(1-3): 135-141.

Marek, M. (1988). *Photosynthetica*, 22:179-183.

Markert, B. (1991). *Vegetatio*, 95:127-135.

Markert, B. (1994). *Element Concentration Cadasters in Ecosystems: Progress report presented at the 25th General Assembly of IUBS* in Paris.

Martinoia, E. (1992). *Botanica Acta*, 105:232-234.

Massimino, D., Andre, D., Richaud, C., Daguenet, A., Massimo, J., and Vivoli, J. (1981). *Physiol. Plant.*, 51:190-195.

Matile, P. (1991). *Veroff Naturf. Ges. Zurich*, 135(5).

Matthies, D. (1995). Parasitic and competitive interactions between the hemiparasites *Rhinanthus serotinus* and *Odontites rubra* and their host *Medicago sativa. J. Ecol.*, 83:245-251.

Meilleur, A., Veronneau, H., and Bouchard, A. (1994). Shrub communities as inhibitors of plant succession in Southern Quebec. *Environ. Mgmt.*, 18(6):907-921.

Menzel, A., and Fabian, P. (1999). *Nature*, 397(6721):659.

Meyer, F.M. (1978). *Baume in der Stadt.* Ulmer, Stuttgart.

Milburn, J.A. (1966). *Planta*, 69:34-42.

Millet, M. (1992). *Bios*, 23(1-2):45-50, 67-69.

Mitscherlich, G. (1970). *Wald, Washtum und Umwelt. Eine Einfuhrung in die Okologischen Grundlagen des Waldwaschstums*, vol. 1, *Form und Wasserhaushalt.* Saverlander, Frankfurt.

Mizutani, J. (1989). Plant allelochemicals and their roles. In Chou, C., and Waller, G.R. (eds.). *Phytochemical Ecology*, Academia Sinica Monograph, Inst. Bot. Ser., Taipei, pp. 155-165.

Mohr, H., and Schopfer, P. (1995). *Plant Physiology.* Springer.

Montieth, J.L. (1972). *J. Appl. Ecol.*, 9:747-766.

Morozov, V.L., and Belaya, G.A. (1988). *Ekologiya dalvenostochnogo Krupnotravya.* Nauka Moskva.

Moujir, L., Angel, M., Gutierrez-Navarro, San Andre, L., and Javier, G.L. (1993). Structure-antimicrobial activity relationships of abietane diterpenes from *Salvia* species. *Phytochemistry*, 34(6):1493-1495.

Moustakas, M., Ouzounidou, G., Eleftheriou, E.P., and Lannoyer (1996). *Plant Physiol. Biochem.*, 34(4):553-560.

Mtolera, M.S.P., Collen, J., Pedesen, M., and Semesi, A.K. (1995). *Eur. J. Phycol.*, 30:289-297.

Mulkey, S.S., and Pearcy, R.W. (1992). *Funct. Ecol.*, 6:719-729.

Muller, C.H. (1965). *Bull. Torrey Bot. Club*, 92(1):38-45.

Muller-Stoll, W.R. (1935). *Z. Bot.*, 29:161-253.

Munns, R., and Sharp, R.E. (1993). *Austr. J. Plant Physiol.*, 20:425-437.

Murphy, S.D., and Aarssen, L.W. (1995). Reduced seed set in *Elytrigia repens* caused by allelopathic pollen from *Plheum pratense. Can. J. Bot.*, 73:1417-1422.

Naessens, M. (1998). Doct. thesis, Ecole Nationale Superieure des Mines de Saint-Etienne and Universite Joseph Fournier de Grenoble.

Ndungu, M., Lwande, W., Hassanali, A., Moreka, L., Chhabra, Sumesh Chander (1995). *Cleome monophylla* essential oil: its constituent as tick (*Rhipicephalus appendiculattus*) and maize weevil (*Sitophilus zeamais*) repellents. *Entomol. Exp. Appl.*, 76:217-222.

Nelson, N.D. (1984). *Photosynthetica*, 18:600-605.

Nilsen, K.N. (1971). *Hort. Sci.*, 6:26-29.

Nobel, D.S. (1977). *Oecologia*, 27:117-133.

Nobel, P.S. (1988). *Environmental Biology of Agaves and Cacti.* Cambridge Univ. Press, Cambridge.

Nobel, P.S. (1991). *Physicochemical and Environmental Plant Physiology.* Academic Press, New York, p. 635.

Nobel, W., Mayer, T.H., and Kohler, A. (1983). *Z. Wasser Abwasser Forsch,* 16:87-90.

Olivella, C., Biel, C., Vendrell, M. and Save, R. (2000). Hortscience 35(2): 222-225.

O'Toole, J.C., Cruz, R.T., and Singh, T.N. (1979). *Plant Sci. Lett.,* 16:111-114.

Ochi, H. (1962). *Bot. Mag.,* 65:112.

Ogino, K., Ninomiya, I., and Yoshikawa, K. (1986). In Hotta, M. (ed.). *Diversity and Dynamics of Plant Life in Sumatra.* Kyoto Univ.

Ogren, E. (1991). *Planta,* 184:538-544.

Ortiz Lopez, A., Ort, D.R., and Boyer, J.S. (1991). *Plant Physiol.,* 96:1018-1025.

Osmond, C.B., Bjorkman, O., and Anderson, D.J. (1980). *Physiological Processes on Plant Ecology. Toward a Synthesis with* Atriplex. Springer, Berlin, Heidelberg, New York.

Paine, R.T. (1971). *Annu. Rev. Ecol. Syst.,* 2:145-164.

Pare, P.W., Zajicek, J., Ferracini, Vera L., and Melo, I.S. (1993). *Ecology,* 21(6/7):649-653.

Patterson, D.T., and Doke, S.O. (1979). *Plant Cell Physiol.,* 20:177-184.

Pelmont, J. (1993). *Bacteries et Environnement.* PUG Universite Joseph Fournier, Grenoble Sciences.

Peltunen-Sainio, P., and Makela, P. (1995). *Acta Agric. Scand., Sect. B, Soil and Plant Sci.,* 45:32-38.

Pennings, S.C., and Callaway, R.M. (1996). *Ecology,* 77(5):1410-1419.

Pereira, J.S. (1995). In Schulze, E.D., and Caldwell, M.M. (1995). *Ecophysiology of Photosynthesis.* Springer, Berlin, Heidelberg, New York.

Petropoulou, Y., Kyparissis, A., Nikolopoulos, D., and Manetas, Y. (1995). *Physiol. Plant.,* 94:37-44.

Pfanz, H. (1987). Aufnahme und Verteilung von Schwefeldioxid in Pflanzlichen Zellen und Organellen. Auswirkungen auf den Stoffweschel. Dissertation, Universitat Wurzburg.

Pfanz, H. (1995). In Schulze, E.D., and Caldwell, M.M. (1995). *Ecophysiology of Photosynthesis.* Springer, Berlin, Heidelberg, New York, chapter 4.

Pfeffer, P. (1970). *L'Asie.* Librairie Hachette.

Pisek, A., and Kemmitzen, R. (1968). *Flora,* 157.

Pisek, A., and Larcher, W. (1954). *Protoplasma,* 44:30-46.

Pisek, A., and Tranquillini (1951). *Physiol. Plant.,* 4:1-27.

Pitman, M.G., and Luttge, U. (1983). In Lange, O.L., Nobel, P.S., Osmond, C.B., and Ziegler, H. (eds.). *Encyclopedia of Plant Physiology,* vol. 12C. Springer, Berlin, Heidelberg, New York.

Platt, T., Rao, S., and Irwin, B. (1983). *Nature,* 300(5992):702-704.

Polster, H. (1967). Wasserhaushalt. In Lyr, H., Polster H., and Fielder, H.J. (eds.). *Geholz Physiologie.* Fischer, Jena.

Prioul (1982). Limiting factors in photosynthesis—from the chloroplast to the plant canopy. In Nelene, C., and Charlier, M. (eds.). *Trends in Photobiology*. Plenum Publ. Co., pp. 633-643.

Prosser, C.L. (1986). *Adaptational Biology: Molecules to Organisms*. Wiley, New York.

Rabe, R. (1990). Bioindikation von Luftverunreinigergen. In Kreeb, K.H. *Methoden zurpflazenokologie und biondikation*. Fischer, Jena, pp. 275-301.

Ramade, F. (1990). *Elements d'Ecologie. Ecologie Fondamentale*. McGraw-Hill, Paris, 403 pp., revised edition.

Ramball, S. (1992). *Vegetatio*, 99:147-153.

Rambler, M.B., Margulis, L., and Fester, R. (1989). *Global Ecology. Towards a Science of the Biosphere*. Academic Press.

Raven, J.A. (1977). *Adv. Bot. Res.*, 5:154-240.

Raven, J.A. (1992). *J. Phycol.*, 28:133-146.

Raven, J.A. (1995). In Schulze, E.D., and Caldwell, M.M. (1995). *Ecophysiology of Photosynthesis*. Springer, Berlin, Heidelberg, New York.

Rawson, H.M., Turner, N.C., and Begg, J.E. (1978). *Austr. J. Plant Physiol.*, 5:195-209.

Reed, H.C., Tan, S.H., Haapanen, K., Killmon, M., Reed, O.K., and Elliot, N.C. (1995). *J. Chem. Ecol.*, 21(4):407-418.

Reekie, E.G. (1996). The effect of elevated CO_2 on developmental process and its implications for plant-plant interactions. In Korner, C., and Bazzaz, F.A. (eds.). *Carbon Dioxide, Populations and Communities*. Academic Press, chapter 22, pp. 333-346.

Reising, H., and Schreiber U. (1992). *Photosynth. Res.*, 31:227-238.

Richter, H. (1972). *Ber. Dtsch. Bot. Ges.*, 85:341-351.

Robin, C., Hay, M.J.M., Newton, P.C.D., and Greer, D.H. (1994). Effect of light quality (red:far-red ratio) at the apical bud of the main stolon on morphogenesis of *Trifolium repens*, L. *Ann. Bot.*, 74:119-123.

Robin, D., Martin, M., and Haerdi, W. (1995). *Arch. Sci. Geneve*, 48(1):19-28.

Rodin, L., and Bazilevich, N.I. (1967). *Production and Mineral Cycling in Terrestrial Vegetation*. Oliver and Boyd, Edinburg.

Roller (1963). *Wetter Leben*, 15:1-12.

Rose, U.S.R., Lewis, W.J., Tumlison, J.H. (1998). *J. Chem. Ecol.*, 24(2):302-319.

Ryel, R.J., Barnes, P.W., Beyschlag, W., Caldwell, N.N., and Flint, S.D. (1990). *Oecologia*, 82:304-315. In Schulze, E.D., and Caldwell, M.M. (eds.). *Ecophysiology of Photosynthesis*. Springer, Berlin, Heidelberg, New York, chapter 20.

Ryle, G.J., Powell, C.E., and Gordon, A.J. (1985). *Exp. Bot.*, 36:634-643.

Saab, I.N., Sharp, R.E., and Pritchard, J. (1992). *Plant Physiol.*, 99:26-33.

Sakai, A., and Larcher, W. (1987). *Frost Survival of Plants. Response and Adaptation to Freezing Stress*. Springer, Berlin, Heidelberg, New York.

Salisbury, F.B. (1982). *Hort. Rev.*, 4:66-105.

Salisbury, F.B. (1985). Plant adaptations to the light environment. In Kavrin, A., Juntilla, O., and Nilsen, J. (eds.). *Plant Production in the North.* Norwegian Univ. Press, Troms, pp. 43-61.

Saradadevi, K., and Raghavendra, A.S. (1992). *Plant Physiol.*, 99:1232-1237.

Schafer, C., Simper, H., and Hofmann, B. (1992). *Plant Cell Environ.*, 15:343-350.

Scheuermann, R., Biehler, K., Stuhlfauth, T., and Fock, H.P. (1991). *Photosynth. Res.*, 27:189-197.

Schlee, D. (1992). *Okologische Biochemie* 2, Aufl. Springer, Berlin, Heidelberg, New York.

Schmitt, J., and Wulf, R.D. (1993). Light spectral quality, phytochrome and plant competition. *Tree*, 8(2).

Schreiber, U., Bilger, W., and Neubauer, C. (1995). In Schulze, E.D., and Caldwell, M.M. (1995). *Ecophysiology of Photosynthesis.* Springer, Berlin, Heidelberg, New York.

Schubert, R. (1991) (ed.). *Bioindication interrestricken Okosystemen* 2, Aufl. Fischer, Stuttgart.

Schulze, E.D. (1970). *Flora*, 159:177-232.

Schulze, E.D., and Caldwell, M.M. (1995). *Ecophysiology of Photosynthesis.* Springer, Berlin, Heidelberg, New York.

Schulze, E.D., Lange, O.L., and Koch, W. (1972). *Oecologia*, 9:317-340.

Schulze, E.D., and Stik, E. (1990). *Angew Bot.*, 64:225-235.

Schulze, W., Stitt, M., Schulze, E.D., Neuhaus, H.E., and Fichtner, K. (1991). *Plant Physiol.*, 95:890-895.

Schulze, W., and Schulze, E.D. (1995). In Schulze, E.D., and Caldwell, M.M. (1995). *Ecophysiology of Photosynthesis.* Springer, Berlin, Heidelberg, New York.

Schwanz, P., Picon, C., Vivin, P., Dreyer, E., Guehl, J.M., and Polle, A. (1996). *Plant Physiol.*, 110(2):393-402.

Scuderi, L.A. (1993). *Science*, 259:1433-1436.

Seeley, E.J., and Kammereck, R. (1977). *J. Amer. Soc. Hort. Sci.*, 102:731-733.

Seidlova, J., and Sarapetka, B. (1997). *Biologia*, 52(1):85-90.

Selye, H. (1936). *Nature*, 138:32.

Selye, H. (1973). *Amer. Sci.*, 61:693-699.

Semiadi, G., Barry, T.N., Muir, P.D., and Hodgson, J. (1995). *J. Agric. Sci.*, Cambridge, 125:99-107.

Senft, W.H. (1978). *Limnol. Oceanogr.*, 23:709-718.

Shah, N., Smirnoff, N., and Stewart, G.R. (1987). *Physiol. Plant.*, 69:699-703.

Sharp, R.E., Matthew, M.A., and Boyer, J.S. (1984). *Plant Physiol.*, 95:95-101.

Shevtsova, A., Ojala, A., Neuvonen, S., Vieno, M., and Haukioja, E. (1995). Growth and reproduction of dwarf shrubs in a subarctic plant community: annual variation and above-ground interactions with neighbours. *J. Ecol.*, 83:263-275.

Sibly, R.M., and Calow, P. (1986). *Physiological Ecology of Animals.* Blackwell Scientific Publications.

Siebel, H.N., Van Wijk, M., and Blom, C.W.P.M. (1998). *Acta Bot. Neerl.,* 47(2):219-230.

Sims, D.A., and Pearcy, R.W. (1991). *Oecologia,* 86:447-453.

Sitte, P., and Eschbach, S. (1992). *Prog. Bot.,* 53:29-43.

Sonoike, K. (1995). *Plant Cell Physiol.,* 36(5):825-830.

Stadler, J., and Gebauer, G. (1992). *Trees,* 6:236-240.

Stanhill, G. (1970). The water flux in temperate forest: precipitation and evapotranspiration. In Reichle, D.E. (ed.). *Analysis of Temperate Forest Ecosystems.* Springer, Berlin, Heidelberg, New York, pp. 247-256.

Sterner, O. (1995). Toxic terpenoids from higher fungi and their possible role in chemical defence systems. *Crytogamie, Mycol.,* 16(1):47-57.

Stetter, K.O., Fiala, G., Huber, G., Huber, R., and Segerer, A. (1990). *Fems Microbiol. Rev.,* 75:117-124.

Steubing, L. (1976). *Landschaft stadt.,* 8:97-144.

Steubing, L., Haneke, J., Biermann, J., and Gnittke, J. (1989). *Angew Bot.,* 63:361-374.

Stickan, W., and Zhang, X. (1992). *Trees,* 6:96-102.

Stitt, M. (1993). In Schulze, E.D. *Flux Control in Biological Systems.* Academic Press.

Stitt, M. and Schulze E.D. (1995) in Schulze E.D. and Caldwellmm (1995) Ecophysiology of photosynthesis. Springer, Berline, Heidelberg, New York.

Stocker, O. (1929). *Planta,* 7:382-387.

Stocker, O. (1931). *Jahrb. Wiss. Bot.,* 75:494-549.

Stocker, O. (1947). *Naturwissenschaften,* 34:362-371.

Stocker, O. (1970). *Flora,* 159:539-572.

Stocker, O. (1974a). *Flora,* 163:46-88.

Stocker, O. (1974b). *Flora,* 163:480-529.

Stone, W., Idle, D.B., and Brennan, R.M. (1993). *Ann. Bot.,* 72,613-617.

Stoy, V. (1965). *Physiol. Plant.,* 4:1-125.

Strittmatter, G., Janssens, J., Opsomer, C., and Botterman, J. (1995). Inhibition of fungal disease development in plants by engineering controlled cell death. *Bio/Technology,* 13.

Stuart-Chapin III, F., Autumn, K., and Pugnaire, F. (1993). *Amer. Naturalist,* 142:578-592.

Sucoff, E. (1975). Tech. Bull. 303, Forestry Service, 20, Agricultural Experiment Station, Minnesota, pp. 3-49.

Sveshnikova, V.M. (1979). *Dominanty Khasakstankikh Stepei.* Nauka, St. Petersbourg.

Svoboda, J. (1977). Ecology and primary production of raised beach communities. In Bliss, L.C. (ed.). *Truelove Lowland, Devon Island, Canada: A High Arctic Ecosystem.* University of Alberta Press, pp. 185-216.

Svoboda, J., and Hosek, P. (1976). *Arctic Alp. Res.*, 8(4):393-398.

Tahvanainen, J., Julkumen-Totto, R., and Kettunen, J. (1985). *Oecologia*, 60:52-56.

Takahashi, H., and Scott, T.K. (1993). *Plant Cell Environ.*, 16:99-103.

Tandeau de Marsac, N. (1979). Doct. thesis in science, University of Paris.

Tang and Chang (1991). *au: cited in p. 201. Please give complete reference.*

Tani, N. (1978). In Takahashi, K., and Yoshino, M. (eds.), *Climatic Change and Food Production.* Tokyo Univ. Press.

Tarayre, M., Thompson, J.D., Escarre, J., and Linhart, V.B. (1995). *Oecologia*, 101(1):110-118.

Tenhunen, J.D., Stegworlf, R.T.W., and Oberauer, S.F. (1995). In Schulze, E.D., and Caldwell, M.M. (1995). *Ecophysiology of Photosynthesis.* Springer, Berlin, Heidelberg, New York, chapter 21.

Terashima, I., Funayama, S., and Sonoike, K. (1994). *Planta*, 193:300-306.

Terry, N., Waldron, LJ., and Taylor, S.E. (1983). In Dale, J.E., and Milthorpe, F.L. (eds.). *Growth and Functioning of Leaves.* Cambridge Univ. Press.

Teughels, H., Nijs, I., Van Hecke, P., and Impens, I. (1995). Competition in a global change environment: the importance of different plant traits for competitive success. *J. Biogeogr.*, 22:297-305.

Thiele, A., Krause, G.H., and Winter, K. (1998). *Austr. J. Plant Physiol.*, 25(2):189-195.

Tollsten, L., Knudsen, J.T., and Bergstrom, G.L. (1994). Floral scent in generalistic *Angelica* (Apiaceae): an adaptive character? *Ecology*, 22(2):161-169.

Tranquillini, W. (1959). *Planta*, 54:107-129.

Tremmel, D.C., and Bazzaz, F.A. (1995). Plant architecture and allocation in different neighborhoods: implications for competitive success. *Ecology*, 76(1):262-271.

Trebst, A. (1995). In Schulze, E.D., and Caldwell, M.M. (1995). *Ecophysiology of Photosynthesis.* Springer, Berlin, Heidelberg, New York, chapter 2.

Tselniker (Zelniker), Yu. L. (ed.). (1993). *Rost i Gazoobmen CO_2 Ulesnykhderevev.* Nauka, Moscow.

Tsuji, J., Evelyn, P., Douglas, A. Gage, Hammerschmidt, R., and Shauna, C. (1992). Somerville phytoalexin accumulation in *Arabidopsis thaliana* during the hypersensitive reaction to *Pseudomonas syringae* pv. *syringae. Plant Physiol.*, 98:1304-1309.

Tyler, G. (1990). *Bot. Linn. Soc.*, 104:231-253.

Ucko, M., Geresh, S., Simon-Berkovitch, B., and Arad, S. (Malis) (1994). Predation by a dinoflagellate on a red microalga with a cell wall modified by sulfate and nitrate starvation. *Mar. Ecol. Prog. Ser.*, 104:293-298.

Vance, H.O., and Franko, D.A. (1997). *J. Freshwater Biol.*, 12(3):405-409.

Van Staden, J., and Grobbelaar, N. (1995). *Environ. Exp. Bot.*, 35(3):321.

Vannier, G. (1983). In Lebrun, P., Andre, H.M., De Medts, A., Gregoire-Wibo, C., and Wauthy, G. (eds.). *New Trends in Soil Biology.* Diev-Brichart, Louvain.

Viollon, C., and Chaumont, J.P. (1994). *Mycopathologia*, 128:151-153.

Vogeli, U., Vogeli-Lange, R., and Chapell, J. (1992). *Plant Physiol.*, 100(3):1369-1376.

Vozzo, S.F., Miller, J.E., Pursley, W.A., and Heagle, A.S. (1995). *J. Environ. Qual.*, 24:663-670.

Vrieling, K. (1991). Cost assessment of the production of pyrrolizidine alkaloids by *Senecio jacobaea* L. II. The generative phase. *Med. Fac. Landbouww. Rijksuniv.*, Ghent, 56/3a.

Vrieling, K., and Van Wijk, C.A.M. (1994a). Cost assessment of the production of pyrrolizidine alkaloids in ragwort (*Senecio jacobaea* L.). *Oecologia*, 97:541-546.

Vrieling, K., and Van Wijk, C.A.M. (1994b). Estimating costs and benefits of the pyrrolizidine alkaloids of *Senecio jacobaea* under natural conditions. *Oikos*, 79:449-454, Copenhagen.

Wang, J., Ives, N.N.E., and Lechowicz, M.J. (1992). *Funct. Ecol.*, 6:469-475.

Wasternach, C., Atzorn, R., Blume, B., Leoppold, J., and Parthier, B. (1994). *Phytochemistry*, 35(1):49-54.

Weiblen, G.D., and Thompson, J.D. (1995). *Oecologia*, 102:211-219.

Weigel, H.J., Halbwachs, G., and Jager, H.J. (1989). *Z. Pflanzenkr. Pflanzenshutz*, 96:203-217.

Weintraub, J.D. and Scoble, M.J. (1995). Lithine moths on ferns: A phylogenetic study of insect-plant interactions. *Biol. J. Linn. Soc.*, 55:239-250.

Weisner, S.E.B. (1993). Long-term competitive displacement of *Typha latifolia* by *Typha angustifolia* in a eutrophic lake. *Oecologia*, 94:451-456.

Weiss, E., and Berry, J.A. (1987). *Biochim. Biophys. Acta*, 894:198-208.

Wilen, R.W., Bruce, E.E., and Gusta, L.V. (1993). *Can. J. Bot.*, 72:1009-1017.

Wimmer, R., and Halbwachs, G. (1992). *Holz. Roh. Werkstoff*, 50:261-267.

Wiskich, J.T., Bryce, J.H., Day, D.A., and Dry, I.B. (1990). *Plant Physiol.*, 93:611-616.

Wolters, V., and Stickan, W. (1991). *Oecologia*, 88:125-131.

Woodrow, I.E., Murphy, D.J., and Latzkoe, J. (1984). *J. Biol. Chem.*, 259:3791-3795.

Wu, L., Till-Botraud, I., and Torres, A. (1998). *New Phytol.*, 107:627-631.

Wullschleger, S.D., Norby, R.J., and Gunderson, C.A. (1992). *New Phytol.*, 121:515-523.

Yamamoto, Y. (1995). *J. Chem. Ecol.*, 21(9):1365-1373.

Yoshie, F. (1989). *Bull. Assoc. Nat. Sci. Sensho Univ.*, 20:75-87.

Zhang, J., and Davies, W.J. (1987) *J. exp. Bot.*, 38: 649-659.

Zhang, J., and Davies, W.J. (1990). *Plant Cell Environ.*, 13:277-285.

Zimmermann, M.H., and Brown, C.L. (1974). *Trees, Structure and Function*, 2nd ed. Springer, Berlin, Heidelberg, New York.

Index